Instructor's Manual

to accompany

Tortora

Introduction to the Human Body

Fourth Edition

William C. Kleinelp, Jr.
Middlesex County College

An imprint of Addison Wesley Longman, Inc.

New York • Reading, Massachusetts • Menlo Park, California • Harlow, England
Don Mills, Ontario • Sydney • Mexico City • Madrid • Amsterdam

Instructor's Manual
to accompany
Tortora *Introduction to the Human Body,*
Fourth Edition

ISBN: 0-673-97533-9

97 98 99 00 01 9 8 7 6 5 4 3 2 1

TABLE OF CONTENTS

CHAPTER CONTENTS

PREFACE

This manual is for professional educators who use Introduction to the Human Body. It is designed to assist in the preparation of lecture material based on the content of the textbook, in the enhancement of textbook coverage by inclusion of additional material, in the preparation of questions for students to answer, and in the selection of resource materials and computer software.

Each chapter in this manual is divided into six major sections:

CHAPTER AT A GLANCE

This section provides, in bullet form, the areas and subareas discussed in each chapter.

CHAPTER SYNOPSIS

The chapter synopsis provides a concise resume of the contents of each chapter in the text.

LEARNING GOALS/STUDENT OBJECTIVES

The goals and objectives are taken from the beginning of each chapter in the text, and are intended as a guideline for student comprehension.

SAMPLE LECTURE OUTLINE

A comprehensive, detailed lecture outline is presented in the same order as the material in the text chapter. Each instructor has his or her own style of presenting lectures. This outline reflects only one method of structuring the materials in the chapter. Important points, terms, and items of interest have been SET OUT IN CAPITALS throughout the lecture outline.

TEACHING TIPS AND SUGGESTIONS

This section is broken down into several subheadings:

A. OVERHEAD TRANSPARENCIES

This listing initially presents the instructor with the related transparencies from Principles of Anatomy and Physiology (PAP), Eighth Edition. The transparency numbers are given as reference. In addition, other sources for transparencies are provided.

B. HELPFUL HINTS

This area provides some information which may make the introduction or the flow of the material easier. It incorporates suggestions for both lecture and lab components.

C. ESSAYS

The essential questions presented in this section may be assigned as written essay questions, or can be used for discussion. In some instances, these essay questions will reach beyond the text material to challenge the student to apply what he or she has learned.

D. TOPICS FOR DISCUSSION

In this section, one or more topics have been suggested for class discussion. Most are topics that encourage classroom participation in areas of current interest.

AUDIOVISUAL MATERIALS

This section is divided into several areas which include videocassettes, 16 mm film, 35 mm (2x2) transparencies, overhead transparencies, and available computer software. The listing contains suggestions which may aid in the reinforcement and supplementation of lecture or laboratory material. The suppliers, addresses, and phone numbers for these resources are given in the following abbreviated form: CARO (Carolina Biological Supply).

William C. Kleinelp, Jr.

SUPPLIERS AND DISTRIBUTORS

There are many types of audiovisual material, models, computer software, and other equipment available to accompany lectures and discussions of appropriate subject matter. Suggestions for some of the materials are included under AUDIOVISUAL MATERIALS at the end of each chapter presentation. Sources for these items are encoded. Not all of the distributors or suppliers are listed, but there are an adequate number to allow the instructor to make a suitable selection.

It is strongly recommended that the instructors preview audiovisual materials before class presentation. This will provide the opportunity to consider the timing of the showing within the framework of the lecture and reflect upon new concepts that should be reinforced, refuted, or discussed. The date of each publication is listed, when available, as well as the length of the presentation. Efforts have been made to provide as many listings as possible, with the majority falling under 30 minutes. The directory which follows provides names, addresses, telephone and toll-free numbers where available. Most suppliers provide free catalogues.

CODE DESIGNATION AND SOURCE

ABB Abbott Laboratories, Audio-Visual Services, Dept. 383, Abbott Park, IL 60064; [312] 937-3933.

ASC American Cancer Society, 219 East 45th Street, New York, NY 10017.

AIMS AIMS Media, Inc., 6901 Woodley Ave., Lincolnwood, IL 60643; [312] 674-2122.

CARLE Carle Medical Communications, 510 W. Main, Urbana, IL 61801; [217] 384-4838.

CARO Carolina Biological Supply, 2700 York Rd., Burlington, N.C. 27215; [919] 584-0381, [800] 334-5551.

CARSL Carosel Films, 260 5th Ave., Room 705, New York, NY 10001; [212] 683-1660.

CBS CBS Special Products, 51 West 52nd Street, New York, NY 10001; [212] 975-5073.

CHUR Churchill Films, 12210 Nebraska Ave., Los Angeles, CA 90025; [213] 207-6600.

COR Coronet/MTI Films and Video, 108 Wilmot Rd., Deerfield, IL 60015-9990; [312] 940-1260.

CRM CRM/McGraw-Hill Films, 674 Via de la Valle, P.O. Box 641, Del Mar, CA 92014; [619] 453-5000.

EBEC Encyclopedia Brittanica Educational Corporation, 425 North Michigan Ave., Chicago, IL 60611; [312] 321-6800.

EFL Ethen Films Ltd., England, Distributed by EBEC.

EIL Educational Images Ltd., P.O. Box 3456, Elmira, NY 14905; [607] 732-1090.

FAD F.A. Davis Co., 1915 Arch St., Philadelphia, PA 19103.

FHS Films for Humanities and Sciences, Inc., P.O. Box 2053, 743 Alexander Rd., Princeton, NJ 08540; [609] 452-1128.

GAF GAF Corporation, 140 West 51st Street, New York, NY 10020.

H&B Halas and Batchelor Cartoon Film Ltd., England, Distributed by EBEC.

H&R Harper*Collins* Publishers, 10 East 53rd Street, New York, NY 10022; [212] 207-7000.

HRM Human Relations Media, 175 Tompkins Ave., Pleasantville, NY 10570; [914] 769-6900.

HSC Hubbard Scientific Corp., 1946 Raymond Dr., P.O. Box 104, Northbrook, IL 60062; [312] 272-7810.

IBIS Ibis Media, 175 Tompkins Ave., Pleasantville, NY 10570; [914] 747-0177.

ICIA ICI America Inc., Concord Pike and Murphy Rd., Wilmingtion, DE 19899.

K&E Keuffel and Esser Co., Educational Media Division, 20 Whippany Rd, Morristown, NJ 07960.

KSU Kent State University, Audio-Visual Services, Kent, OH 44242; [216] 672-3456.

LILLY Eli Lilly and Company, Medical Division, Indianapolis, IN 46306.

McG CRM/McGraw-Hill Films, 674 Via de la Valle, P.O. Box 641, Del Mar, CA 92014; [619] 453-5000.

MG Media Guild, 11722 Sorrento Valley Rd., Suite E, San Diego, CA 92121; [619] 755-9191.

NBC NBC Educational Enterprises, 30 Rockefeller Plaza, New York, NY 10112; [212] 664-3737.

NSC Nova Scientific Corporation, 2990 Anthony Rd., P.O. Box 50, Burlington, NC 27215.

PFI Polymorph Films Inc., 331 Newbury Sty., Boston, MA 02115.

PLP Projected Learning Programs Inc., P.O. Box 3008, Paradise, CA 95969; [916] 893-4223.

PRM Peter M. Robeck and Co., Distributed by Time-Life.

PSP Popular Science Publishing Co., 355 Lexington Ave., New York, NY 10017.

PSU Pennsylvania State University Audio-Visual Services, Special Services Building, University Pk, PA 16802; [814] 865-6314.

PYR Pyramid Films and Video, Box 1048, Santa Monica, CA 90406; [213] 828-7577.

RJB Robert J. Brady Co., 130 Q Street, NE, Washington, DC 20002.

SKF Smith, Kline, and French Laboratories, Services Dept., 1530 Spring Garden St., Philadelphia, PA 19101.

SM Science and Mankind Inc., Box 2000, 90 South Bedford Rd., Mt. Kisco, NY 10549; [914] 666-4100.

TGC The Graphic Curriculum, P.O. Box 5651, Lenox Hill Station, New York, NY 10021.

TLF Time-Life Films Inc., 43 West 16th Street, New York, NY 10001.

TLV Time-Life Video, 1271 Avenue of the Americas, New York, NY 10020. Distribution is P.O. Box 644, Paramus, NJ 07653; [201] 843-4545.

UCEMC University of California Extension Media Center, 2176 Shattuck Ave., Berkeley, CA 94704; [415] 642-0460.

UIFC University Of Illinois Film Center, 1325 South Oak St., Champaign, IL 61820; [217] 333-1360.

UMedia University Media, Distributed by MG.

USNAC U.S. Audio-Visual Center, GSA Building, Washington, DC 20409; [301] 763-1896.

WNSE Wards Natural Science Establishment Inc., P.O. Box 1712, Rochester, NY 14622.

TABLES

The following represents tables which are listed in the text of this instructor's manual. Additional tables, charts, exhibits, and illustrations can be found in both the accompanying textbook as well as *Principles of Anatomy and Physiology,* by Tortora & Grabowski, Eighth Edition.

ACKNOWLEDGMENTS

I wish to thank Bonnie Roesch for again having the confidence in me to write this instructor's manual. I also wish to thank Cyndy Taylor for her excellent work as the developmental editor. Lastly, I wish to thank Stefanie Schwalb for her excellent editing skills, finding commas where commas should not go.

CHAPTER 1

ORGANIZATION OF THE HUMAN BODY

CHAPTER AT A GLANCE

- ANATOMY AND PHYSIOLOGY DEFINED
- LEVELS OF STRUCTURAL ORGANIZATION
- HOW BODY SYSTEMS WORK TOGETHER
- LIFE PROCESSES
- HOMEOSTASIS: MAINTAINING PHYSIOLOGICAL LIMITS
- *Stress and Homeostasis*
- *Homeostasis of Blood Pressure (BP)*
- ANATOMICAL POSITION
- DIRECTIONAL TERMS
- PLANES AND SECTIONS
- BODY CAVITIES
- ABDOMINOPELVIC REGIONS AND QUADRANTS
- WELLNESS FOCUS: LIFESTYLE, HEALTH, AND HOMEOSTASIS

I. CHAPTER SYNOPSIS

The first chapter gives the student an overview of the general organization of the human body. It presents the definitions of anatomy and physiology, and demonstrates the relationship between structure and function. The students are introduced to the structural levels of the body beginning with the simplest units, atoms and molecules, and progress through cells, tissues, organs, and systems. The general functions of the principal body systems are also described along with the basic life processes and structural plan of the entire body. Characteristics of anatomical position, anatomical names for common body regions, directional terms, planes and sections of the body, body cavities, abdominopelvic regions and quadrants are among the topics covered. Considerable emphasis is placed on homeostasis, stress, and feedback systems. The role of the endocrine and nervous systems in controlling homeostasis is examined, while the regulation of blood pressure and blood sugar levels are used as examples to demonstrate the feedback system.

II. LEARNING GOALS/STUDENT OBJECTIVES

1. Describe the levels of structural organization that compose the human body.
2. Briefly explain how body systems relate to each other.
3. Define the life processes of humans.
4. Define homeostasis and describe its importance in health and disease.
5. Describe the characteristics of anatomical position.
6. Describe several planes that may be passed through the human body and explain how sections are made.

III. SAMPLE LECTURE OUTLINE

A. INTRODUCTION

1. Define ANATOMY as the study of the structure and shape of the body and its parts.
2. Define PHYSIOLOGY and how the body and its parts function, and emphasize that the structure of a part determines how it will function.
3. State that most functions of the body occur to maintain HOMEOSTASIS.
4. Define homeostasis as the ability of the body to maintain a constant internal environment (e.g. constant body temperature, normal blood pH, etc.)
5. Explain in basic terms the relationship between the body's homeostasis, sickness, aging, and death.

B. LEVELS OF STRUCTURAL ORGANIZATION

1. The lowest level of organization is the CHEMICAL LEVEL, comprised of atoms, forming molecules and compounds.
2. Molecules combine to form the CELLULAR LEVEL, which is the basic structural and functional unit of an organism.
3. Groups of similar cells that together perform a particular function represent the TISSUE LEVEL OF ORGANIZATION. This level is represented by four basic tissue types: epithelial tissue, connective tissue, muscle tissue, and nervous tissue.
4. When different types of tissue join together they form the ORGAN LEVEL. These structures are composed of two or more different tissues, have specific functions, and usually have a recognizable shape.
5. The grouping of related organs that have a common function forms the SYSTEM LEVEL.
6. The highest level of organization is the ORGANISMIC LEVEL, which includes all systems of the body that are united to form the ORGANISM.

C. HOW BODY SYSTEMS WORK TOGETHER

1. Body systems work together to maintain health, protect one from disease, and allow for the reproduction of the species.
2. The integumentary system, for example, protects all systems by serving as a barrier between the outside environment and internal tissues and organs.
3. The skin also produces Vitamin D.

D. LIFE PROCESSES

1. METABOLISM is the sum of all chemical reactions in the body. The two major divisions are: CATABOLISM, providing the energy needed to sustain life through the breaking down of food materials; and ANABOLISM, which uses the energy from catabolism to make numerous substances.
2. RESPONSIVENESS is the ability to detect and respond to changes in the internal and/or external environment.
3. MOVEMENT includes motion of the whole body, individual organs, single cells, or even structures within cells.

4. GROWTH refers to the increase in size of existing cells, the number of cells, or the amount of substance surrounding cells.
5. DIFFERENTIATION is the process whereby unspecialized cells become specialized.
6. REPRODUCTION refers to the formation of new cells for growth, repair or replacement, or the production of a new individual.

E. HOMEOSTASIS: MAINTAINING PHYSIOLOGICAL LIMITS

1. HOMEOSTASIS is a steady state or equilibrium. All body systems attempt to maintain homeostasis.
2. Homeostasis is controlled mainly by the nervous and endocrine systems.
3. A stress is any stimulus that attempts to disrupt homeostasis.
4. The body can act to counter the effects of stress through the use of feedback mechanisms.
5. A negative feedback system is one in which the reaction of the body (output) counteracts the stress (input) in order to maintain homeostasis. Most feedback systems in the body are negative.
6. A positive feedback system is one in which the output intensifies the input. If positive feedback should occur, its results are usually destructive or deadly. However, a few are beneficial such as contractions in childbirth.

F. STRESS AND HOMEOSTASIS

1. STRESS is any stimulus that creates an imbalance in the internal environment.
2. Stress may originate from either internal or external stimuli.
3. Stress is counteracted by the regulating or homeostatic devices that bring the internal environment back into balance, and are under the control of the NERVOUS and ENDOCRINE SYSTEMS.
4. The NERVOUS SYSTEM regulates homeostasis by detecting when the body deviates from the balanced state and sending impulses to the proper organs to counteract stress.
5. The ENDOCRINE SYSTEM is a group of glands that secrete chemical messengers called HORMONES, into the blood.

G. HOMEOSTASIS OF BLOOD PRESSURE (BP)

1. BLOOD PRESSURE is the force of blood as it passes through the vessels. In order to sustain life it must be maintained at an appropriate pressure.
2. High blood pressure contributes to the development of heart attacks and stroke.
3. Blood pressure depends on the rate and strength of the heartbeat. If the heart beats faster, more blood pushes into the arteries and elevates the pressure.
4. Increased pressure is detected in the nerve ending in the walls of certain blood vessels. These respond by sending impulses to the brain.
5. The brain, in response, sends nerve impulses to the heart and certain blood vessels to slow the rate, thus decreasing pressure.
6. This cycle is called a FEEDBACK SYSTEM.
7. A FEEDBACK SYSTEM involves a cycle of events in which the information about body conditions is continually monitored and fed back to a central control region.
8. The components of a feedback system are a CONTROL CENTER, a RECEPTOR, and an EFFECTOR.

9. A NEGATIVE FEEDBACK reverses the original condition. A POSITIVE FEEDBACK enhances the original stimulus.

H. TERMINOLOGY

1. Directional terms are always given in regard to the body in ANATOMICAL POSITION.
2. When the body is in correct anatomical position, the body is erect and facing forward, the upper extremities are at the sides with the palms of the hands facing forward, and the feet are flat on the floor.
3. DIRECTIONAL TERMS are used to indicate the relationship of one body part to another.
4. Commonly used terms include but are not limited to:

- SUPERIOR- above or towards the head
- INFERIOR- below or towards the feet
- ANTERIOR/VENTRAL- towards the front of the body
- POSTERIOR/DORSAL- towards the back of the body
- MEDIAL- towards the body midline
- LATERAL- away from the body midline
- INTERMEDIATE- between lateral and medial structures
- PROXIMAL- nearer to the point of attachment of an extremity to the trunk or a structure
- DISTAL- farther from the point of attachment of an extremity to the trunk or a structure
- SUPERFICIAL- on or near the body surface
- DEEP- inward or away from the surface of the body

I. BODY PLANES

1. BODY PLANES or sections are imaginary flat surfaces that are used to divide the body into definite areas.
2. The SAGITTAL PLANE is a vertical plane that divides the body into right and left parts. If the right and left parts are equal in size, it is called a MIDSAGITTAL plane and runs through the midline of the body. If the body is divided into unequal right and left sides, the plane is called a PARASAGITTAL plane.
3. The FRONTAL (CORONAL) PLANE divides the body into anterior and posterior portions. The HORIZONTAL (TRANSVERSE) plane divides the body into superior and inferior portions.

J. BODY CAVITIES

1. CAVITIES are spaces in the body that contain internal organs.
2. The two principal cavities are the DORSAL and VENTRAL CAVITIES.
3. The DORSAL CAVITY is subdivided into the CRANIAL CAVITY, which is in the space within the skull containing the brain, and the SPINAL CAVITY which extends from the cranial cavity through the vertebral column and contains the spinal cord.

4. The VENTRAL CAVITY is subdivided by the diaphragm into the THORACIC CAVITY and MEDIASTINAL CAVITY which contain the heart, lungs, associated blood vessels, and respiratory organs, and the ABDOMINOPELVIC CAVITY which lies inferior to the diaphragm.

K. ABDOMINOPELVIC REGIONS AND QUADRANTS

1. The FOUR QUADRANTS of the abdominopelvic cavity are the right upper quadrant (RUQ), the left upper quadrant (LUQ), right lower quadrant (RLQ), and the left lower quadrant (LLQ).
2. The names of the nine ABDOMINOPELVIC REGIONS are epigastric, right hypochondriac, left hypochondriac, umbilical, right lumbar, left lumbar, hypogastric (pubic), right iliac (inguinal), and left iliac (inguinal).

IV. TEACHING TIPS AND SUGGESTIONS

A. HELPFUL HINTS

1. Encourage memorization of new terms and definitions, especially those which will be used continuously throughout the course.
2. Emphasize that this introductory chapter is providing information that is laying the foundation upon which the rest of the course will be based.
3. Try to relate the information to experiences the students may have in daily life.
4. Use the student's own body whenever possible to show levels of organization and to demonstrate terminology.
5. Different planes and sections can be illustrated by cutting a piece of fruit appropriately.
6. Use the examples of cruise control in a car or a home thermostat to demonstrate negative feedback control.
7. Have the student search literature, newspapers, and magazines for articles and information pertaining to anatomy and physiology.

B. ESSAY QUESTIONS

1. If you were to make a cross section of the body through the diaphragm, which body cavities would you separate? Name several organs that you would be able to see in the upper and lower cavities. Would you be able to see the mediastinum? Explain your answer.
2. Using the terms superior, inferior, anterior, posterior, medial, and lateral, describe the position of the stomach relative to other organs in the abdominal cavity.
3. Describe in detail how the operations of a household thermostat and a negative feedback system are the same.
4. Are any positive feedback systems beneficial? How? Give examples.

C. TOPICS FOR DISCUSSION

1. Discuss the importance of anatomical position. Are there any oddities?
2. Discuss the effects of stress in daily life and stress management.

V. AUDIOVISUAL MATERIALS

A. Overhead Transparencies

1. Principles of Anatomy and Physiology (PAP), 8/e Transparency set (Trs. 1.1, 1.2-1.5a&b, 1.6, 1.7a, 1.9b, 1.11a & 1.12).
2. Human Anatomy: Biology (2 Units; C; EBEC).

B. Videocassettes

1. Homeostasis: Maintaining the Stability of Life (36 min.; 1988, CFM).
2. The Million Dollar Scan (30 min.; 1984; LCA/KSU).
3. Stress: Is Your Lifestyle Killing You? (29 min.; 1985; KSU).
4. Stress: Learning to Handle It (23 min.; 1984; KSU/SIF).

C. Films: 16 mm

1. The Human Body: Systems Working Together (15 min.; 1980; COR/KSU).
2. Man: The Incredible Machine (28 min.; 1975; NGF).
3. Incredible Voyage (26 min.; 1968; McG).
4. Basic Anatomy and Physiology of the Mammal: Introduction to Dissection (5 min.; UIFC).

D. Transparencies: 35 mm (2x2)

1. Principles of Anatomy and Physiology (PAP), Slide Set (# 1-8).
2. Atlas of Human Anatomy (AHA), Slide set.
3. Topographical Anatomy (Slides 206-224; McG).

CHAPTER 2

INTRODUCTORY CHEMISTRY

CHAPTER AT A GLANCE

I. CHAPTER SYNOPSIS

This chapter provides students with the essential chemical background needed to understand the physiology of the body. Among the topics considered are chemical elements, atomic structure, molecule formation, ionic, covalent and hydrogen bonding, and chemical reactions. The structure and importance of the inorganic substances such as water, acids, bases, and salts are emphasized. The structure and importance of the organic substances described include carbohydrates, lipids, proteins, nucleic acids (DNA and RNA), ATP, and cyclic AMP. The concept of pH and the role of buffer systems in maintaining homeostasis are also considered.

II. LEARNING GOALS/STUDENT OBJECTIVES

1. Identify the principal chemicals found in the human body.
2. Describe the structure of the atom.
3. Explain how chemical bonds form.
4. Define a chemical reaction and explain why it is important to the human body.
5. Discuss the functions of water and inorganic acids, bases, and salts.
6. Define pH and explain how the body attempts to keep pH within the limits of homeostasis.
7. Discuss the functions of carbohydrates, lipids, and proteins.
8. Explain the importance of deoxyribonucleic acid (DNA), ribonucleic acid (RNA), and adenosine triphosphate (ATP).

III. SAMPLE LECTURE OUTLINE

A. CHEMICAL ELEMENTS

1. MATTER is anything that occupies space and has mass. It is made up of building units called atoms.
2. An ELEMENT is a substance which cannot be broken down into a simpler substance by ordinary chemical means.
3. Oxygen, carbon, hydrogen, and nitrogen make up 96% of the body's weight. These elements together with calcium and phosphorous make up 99% of the body's weight.
4. An ATOM is the smallest part of an element with the same physical properties of that element.

B. STRUCTURE OF AN ATOM

1. Atoms consist of a NUCLEUS which contains PROTONS and NEUTRONS. The protons are positively charged while the neutrons have no charge, or are "neutral."
2. An atom also has negatively charged ELECTRONS that travel randomly around the nucleus in specific volumes of space called orbitals.
3. The total number of protons in an atom equals its ATOMIC NUMBER. This number is equal to the number of electrons in an electrically neutral atom.
4. The combined total of protons and neutrons of an atom equals its ATOMIC MASS.
5. It is the outer shell electrons that participate in a chemical reaction and are referred to as VALENCE ELECTRONS.

C. MOLECULES

1. A MOLECULE consists of two or more atoms joined by chemical bonds. A COMPOUND is a molecule or group of molecules containing more than one type of atom.
2. Two main types of chemical bonds are IONIC BONDS and COVALENT BONDS.
3. In an IONIC bond, outer-energy-level electrons are transferred from one atom to another. The transfer forms ions, whose unlike charges attract one another and form ionic bonds.
4. In a COVALENT BOND, there is a sharing of pairs of outer-energy-level electrons.
5. HYDROGEN BONDS are found between water molecules and large complex organic molecules.
6. Hydrogen bonding provides temporary bonding between certain atoms within large complex molecules such as proteins and nucleic acids.

D. CHEMICAL REACTIONS

1. A chemical reaction is the process by which the bonds between atoms are rearranged to form new substances.
2. SYNTHESIS REACTIONS produce a new molecule. The reactions are anabolic; bonds are formed.
3. In DECOMPOSITION REACTIONS, a substance breaks down into other substances; the reactions are CATABOLIC and bonds are broken.
4. When chemical bonds are formed, energy is needed; when bonds are broken, energy is released.

E. INORGANIC COMPOUNDS

1. Inorganic substances usually lack carbon, contain ionic bonds, resist decomposition, and dissolve readily in water.
2. Examples of inorganic compounds important to the body include: water, carbon dioxide, oxygen, and ammonia.
3. Water is the most abundant and important substance in the body. It is an excellent solvent and suspending medium, participates in chemical reactions, absorbs and releases heat slowly, and lubricates. Water is involved in digestion, the elimination of wastes, circulation, and the regulation of body temperature.
4. Inorganic acids, bases, and salts dissociate into ions in water. Cations are positively charged ions; anions are negatively charged ions. An acid ionizes into H^+ ions and a base ionizes into OH^- ions.
5. The pH of different parts of the body must remain fairly constant for the body to remain healthy. On the pH SCALE, 7 represents neutrality. Values below 7 indicate ACID solutions, and values above 7 represent BASIC solutions.

6. The pH values of different parts of the body are maintained by buffer systems which consist of weak acids and bases. Buffer systems eliminate excess hydrogen and hydroxyl ions in order to maintain pH homeostasis.

F. ORGANIC COMPOUNDS

1. Organic substances always contain carbon and hydrogen. Most organic substances contain covalent bonds and many are insoluble in water.
2. CARBOHYDRATES are sugars or starches, and are the most common sources of energy needed for life. They may be monosaccharides, dissacharides, or polysaccharides. Carbohydrates, and other organic molecules, are joined together to form larger molecules through a process called DEHYDRATION SYNTHESIS. In the reverse process, called digestion, HYDROLYSIS occurs in order to break down large molecules, with the addition of water.
3. LIPIDS are a diverse group of compounds that include fats, phospholipids, steroids, carotenes, Vitamins E and K, and prostaglandins (PGs). Fats protect, insulate, and provide energy.
4. PROTEINS are constructed from amino acids. They give structure to the body, regulate enzyme processes, provide protection, and help to contract muscle.
5. DNA and RNA are nucleic acids consisting of nitrogenous bases, sugar and phosphate groups. DNA is a double helix and is the primary chemical in genes. RNA differs in

structure and chemical composition from DNA and is mainly concerned with protein synthesis reactions.

6. The principal energy-storing molecule in the body is ATP. ATP decomposes to yield ADP and energy. ATP is manufactured from ADP and phosphate, primarily by using the energy supplied by the decomposition of glucose in a process called cellular respiration.
7. Cellular respiration may be ANAEROBIC (without molecular oxygen) or AEROBIC (in the presence of molecular oxygen). Aerobic respiration provides energy to generate ATP.

IV. TEACHING TIPS AND SUGGESTIONS

A. Helpful Hints

1. Give each student their own periodic table and have them practice locating the symbol of the element, atomic number, atomic mass, number of electrons in the outer shell, and so on.
2. Use a toothpick with styrofoam balls to demonstrate bonding between atoms and to illustrate the various types of chemical reactions. A molecular model kit should be used.
3. Use the "DNA made EASY" kit (#17-1040), which can be purchased from Carolina Biological supply. This kit is an excellent way to demonstrate the structure of DNA and RNA. It can also be used to illustrate the processes of DNA replication, RNA transcription, and protein synthesis.
4. Perform simple chemical tests to demonstrate and detect the presence of various organic molecules in certain foods with indicators such as iodine, Biuret solution, and Benedict's reagent.

B. Essay Question

1. The formation of ionic bonds involves the transfer of outer-energy-level electrons from one atom to another. Draw a series of diagrams to illustrate ionic bond formation in a molecule of magnesium chloride.
2. Buffer systems are vital chemicals that help maintain the pH of the body fluids. Using the carbonic acid/bicarbonate buffer pair as an example, explain how the buffer operates to counteract a strong acid and a strong base.
3. Explain how DNA is ultimately responsible for the formation of proteins by incorporating the processes of RNA transcription and RNA translation.

C. Topics for Discussion

1. Why is chemistry important to the study of anatomy and physiology?
2. Discuss how the body would attempt to maintain proper blood pH if a small quantity of hydrochloric acid were ingested.
3. Describe the importance of ATP.
4. Identify some genetically engineered products presently available.

V. AUDIOVISUAL MATERIALS

A. Overhead Transparencies

1. PAP Transparency Set, 8/e (Trs. 2.1, 2.6a-c, 2.7, 2.10a-c, 2.12, 2.13, 2.14a-d, 2.15 & 2.16a-b).
2 Basis: DNA-RNA Gene Expression (WNSE).
3. The Chemistry of Life (20 Transparencies; BM).

B. Videocassettes

1. Acids, Bases, and Salts (20 min.; 1981; COR/KSU).
2. DNA and RNA: Deciphering the Code of Life (36 min.; 1988; CHF).
3. The Chemistry of Life (EIL).
4. The Chemistry of Proteins (EIL).
5. The Chemistry of Carbohydrates and Lipids (EIL).
6. The Chemistry of Nucleic Acids (EIL).

C. Films: 16 mm

1. Acids, Bases, and Salts (20 min.; 1981; COR/KSU).
2. The Energetics of Life (23 min.; 1972; JW/KSU).
3. Molecular Biology (15 min.; 1981; KSU/COR).
4. The Structure of Protein (16 min.; 1970; BFA/KSU).
5. Chemical Man (8 min.; 1968; PYR).
6. A Drop of Water (15 min.; ACI).
7. Chemistry of the Cell II: Function of DNA and RNA in Protein Synthesis (16 min.; McG).
8. DNA: The Blueprint of Life (18 min.; 1968; JW/KSU).
9. DNA: The Thread of Life (24 min.; 1978; KSU).
10. DNA: Molecule of Heredity (16 min.) 1960; EBEC/KSU).

D. Transparencies: 35mm (2x2)

1. Buffers (64 Slides; CARO).
2. PAP Slide Set (Slide 9).
3. The Chemistry of Life (80 Slides; EIL).
4. The Chemistry of Proteins (66 Slides; EIL).
5. The Chemistry of Carbohydrates and Lipids (67 Slides; EIL).
6. The Chemistry of Nucleic Acids (53 Slides; EIL).

E. Filmstrips

1. Basic Biochemistry (PRM).
2. Chemical of Life (PLP).

F. COMPUTER SOFTWARE

1. Basic Biology Series: Lipids, Proteins, Nucleic Acids, Carbohydrates Biochemistry Test (Apple C4071; C4072; C4073; C4074; C34075; EIL).
2. Biochemistry Series: Proteins, Nucleic Acids, Lipids, Carbohydrates Biochemistry Test (Apple II, R-570075 [set only]; PLP).
3. Chemistry for Biologists (Apple; IBM; C3024; EIL).

CHAPTER 3

CELLS

CHAPTER AT A GLANCE

- GENERALIZED ANIMAL CELL
- PLASMA (CELL) MEMBRANE
- CHEMISTRY AND STRUCTURE
- *Functions*
- *Movement of Materials Across Plasma Membranes*
- *Passive Processes*
- *Active Processes*
- CYTOSOL
- ORGANELLES
- *Nucleus*
- *Endoplasmic Reticulum (ER)*
- *Ribosomes*
- *Golgi Complex*
- *Lysosomes*
- *Mitochondria*
- *Cytoskeleton*
- *Flagella and Cilia*
- *Centrosomes and Centrioles*
- GENE ACTION
- *Protein Synthesis*
- *Transcription*
- *Translation*
- NORMAL CELL DIVISION
- *Somatic Cell Division*
- *Mitosis*
- *Cytokinesis*
- ABNORMAL CELL DIVISION
- *Definition*
- *Growth and Spread*
- *Causes*
- *Treatment*
- MEDICAL TERMINOLOGY AND CONDITIONS
- WELLNESS FOCUS: CANCER-PREVENTION LIFE STYLE

I. CHAPTER SYNOPSIS

Students are presented with the structural and functional aspects of cells through an analysis of a generalized cell. Each cellular component is detailed in terms of structure and function. Important transport mechanisms, including diffusion, osmosis, filtration, dialysis, and active transport are considered. Attention is also given to the events involved in protein synthesis, and the mechanism and importance of somatic and reproductive cell divisions. The chapter concludes with cancer and a list of medical terminology.

II. LEARNING GOALS/STUDENT OBJECTIVES

1. List the parts of the cell.
2. Explain the structure and functions of the plasma membrane.
3. Describe how materials move across plasma membranes.
4. Describe the structure and functions of organelles.
5. Define a gene and explain the sequence of events involved in protein synthesis.
6. Discuss the stages, events, and significance of cell division.
7. Describe cancer as a homeostatic imbalance of cells.

III. SAMPLE LECTURE OUTLINE

A. THE GENERALIZED CELL

1. CYTOLOGY is the study of cells.
2. A CELL is the smallest unit that can exist as an independent structure and contains all the necessary components for life.
3. The cell is the basic structural, functional, and living unit of the body.
4. The two main types of cells are PROKARYOTIC cells and EUKARYOTIC cells. The body cells are eukaryotic.
5. PROKARYOTIC cells are relatively simple cells that lack an organized nucleus and membrane-bound organelles. Bacteria are an example of this type of cell.
6. EUKARYOTIC cells are more complex and contain an organized nucleus with a plasma membrane, cytoplasm, organized membrane-bound organelles, and inclusions.

B. THE PLASMA MEMBRANE

1. The PLASMA MEMBRANE is a living, responding structure which surrounds the cell and is composed primarily of phospholipids and protein.
2. It is SELECTIVELY PERMEABLE, which means that it will allow some substances to pass through while inhibiting the entrance of others, depending on their molecular size, lipid solubility, electrical charge, and the presence of carriers.

C. CHEMISTRY AND STRUCTURE

1. The plasma membrane consists mostly of phospholipids and proteins. In lesser amounts we also find cholesterol, glycolipids, and glycoprotiens.
2. The phospholipids are arranged in two parallel rows forming a PHOSPHOLIPID BILAYER. There are two kinds of proteins. These are the INTEGRAL PROTEINS, which penetrate through the phospholipid bilayer, and PERIPHERAL PROTEINS, which are loosely attached to the exterior or interior of the cell.

3. Some integral proteins form CHANNELS, which contain PORES through which transport substances move into and out of the cell. Others act as TRANSPORTERS, which identify and attach to specific molecules such as hormones.

D. FUNCTION

1. The major functions of the plasma membrane are:
   - COMMUNICATION
   - ELECTROCHEMICAL GRADIENT
   - SHAPE AND PROTECTION
   - SELECTIVE PERMEABILITY
2. The selective permeability of the plasma membrane to different substances depends upon seal factors. These are:
   - LIPID SOLUBILITY
   - CHARGE
   - PRESENCE OF CHANNELS AND TRANSPORTERS

E. MOVEMENT OF MATERIALS ACROSS THE PLASMA MEMBRANE

1. The body fluid inside the cell is called INTRACELLULAR FLUID (ICF). The fluid outside of the cell is called EXTRACELLULAR FLUID (ECF).
2. The fluid found in the microscopic spaces between cells is called the INTRACELLULAR FLUID. The ECF in the blood vessels is called PLASMA, and in the lymphatic vessels it is called LYMPH.
3. The interstitial fluid contains gases needed for maintaining life. The interstitial fluid is also called the BODY'S INTERNAL ENVIRONMENT.
4. The transport processes involved in the movement of materials between cells are PASSIVE TRANSPORT and ACTIVE TRANSPORT.
5. Passive transport utilizes KINETIC ENERGY, which is the energy of molecules in motion. Molecules will move from higher concentrations to lower concentrations until molecules are evenly distributed or EQUILIBRIUM is established.
6. The difference between high and low concentrations is called a CONCENTRATION GRADIENT.

F. PASSIVE PROCESSES

1. PASSIVE PROCESSES involve only the kinetic energy of individual molecules (no ATP is required). Examples of this type of movement are diffusion, osmosis, facilitated diffusion, and filtration.
2. SIMPLE DIFFUSION is the net movement of molecules or ions from an area of higher concentration to an area of lower concentration until dynamic equilibrium is reached.
3. In FACILITATED DIFFUSION, certain molecules, such as glucose, combine with a membrane carrier to become soluble in the phospholipid portion of the membrane.
4. OSMOSIS is the movement of water through a selectively permeable membrane from an area of higher water concentration to an area of lower water concentration until dynamic equilibrium is reached.
5. Osmosis can be understood by considering the effects of different concentrations of water. A solution which has equal concentrations of solute and solvent (water) on either side of the membrane is said to be ISOTONIC. If the solution has a high concentration of solute and a low

concentration of solvent, the solution is HYPERTONIC. If the solution has a low concentration of solute and a high concentration of solvent, the solution is HYPOTONIC.

6. FILTRATION involves the movement of solvents such as water and dissolved substances across a selectively permeable membrane by gravity or mechanical pressure.

G. ACTIVE PROCESSES

1. ACTIVE PROCESSES involve the use of ATP by the cell. Examples are endocytosis, exocytosis, and receptor-mediated endocytosis.
2. ACTIVE TRANSPORT is the movement of ions across a cell membrane from a lower concentration to a higher concentration with the use of ATP.
3. ENDOCYTOSIS is the movement of substances through plasma membrane in which the membrane surrounds the substance, encloses it, and brings it into the cell. This includes phagocytosis, pinocytosis, and receptor-mediated endocytosis.
4. PHAGOCYTOSIS is the ingestion of solid particles by pseudopodia. It is the process white blood cells use to destroy bacteria and other foreign agents.
5. PINOCYTOSIS is the ingestion of liquid by the plasma membrane. In this process, the liquid becomes surrounded by a vacuole.
6. RECEPTOR-MEDIATED ENDOCYTOSIS is the selective uptake of large molecules and particles by the cell.

H. CYTOSOL

1. All the cellular contents between the plasma membrane and nucleus make up the CYTOPLASM. The thick, semi-fluid portion of the cytoplasm is the CYTOSOL.
2. CYTOSOL is the intracellular material inside the cell between the plasma membrane and the nucleus that contains the organelles and inclusions.
3. It is composed mostly of water plus proteins, carbohydrates, lipids, and inorganic substances.
4. Functionally, cytosol is the medium in which some chemical reactions occur.

I. CELLULAR ORGANELLES

1. Cellular organelles can be thought of as little organs. Just as organs function to keep the body alive, organelles are specialized structures that perform specific functions vital to the life of a cell.
2. The NUCLEUS is the most complex structure in the cell. It is involved in the regulation of cellular activities and contains the genetic information (DNA). Most body cells have a single nucleus. Exceptions to this are red blood corpuscles which have no nucleus and skeletal muscle cells which may have several nuclei. The parts of the nucleus include the nuclear membrane, nucleoplasm (containing the chromosomes in chromatin form), and nucleoli (which contain RNA and are thought to form ribosomes).
3. The ENDOPLASMIC RETICULUM (ER) is a network of parallel membranes continuous with the plasma membrane and nuclear membrane. GRANULAR, or rough ER, has ribosomes attached to it. AGRANULAR, or smooth ER, does not contain ribosomes. The ER provides mechanical support, exchanges materials with the cytoplasm, and transports materials within the cell. Granular ER stores the protein synthesized by ribosomes. Agranular ER is involved in the production of lipids, especially steroids.

4. RIBOSOMES are granular structures consisting of ribosomal RNA and ribosomal proteins. They occur free or attached to endoplasmic reticula and are the site of protein synthesis.
5. The GOLGI COMPLEX consists of four to eight stacked, flattened, membranous sacs called cisternae. The principal function of the Golgi is to process, sort, and deliver proteins within the cell. It also secretes lipids and forms lysosomes.
6. LYSOSOMES are spherical structures that contain digestive enzymes. They are formed in the Golgi complexes and are found in large numbers in white blood cells which carry on phagocytosis. Digestion by lysosomes of worn-out cell parts is called AUTOPHAGY; programmed self-destruction by lysosomes is called AUTOLYSIS. Lysosomes also function in extracellular digestion.
7. MITOCHONDRIA consist of a smooth outer membrane and a folded inner membrane surrounding the interior matrix. The inner folds are called CRISTAE. The mitochondria are called the "powerhouses of the cell" because they produce ATP.
8. MICROFILAMENTS, MICROTUBULES, and INTERMEDIATE FILAMENTS form the cytoskeleton. MICROFILAMENTS are rod-like structures consisting of the protein actin or myosin. They are involved in muscular contraction, support, and movement. MICROTUBULES are slender tubes made of the protein tubulin. They provide support, movement, form flagella, cilia, centrioles and the mitotic spindle, and intermediate filaments appear to provide structural reinforcement in some cells.
9. CENTRIOLES are paired cylinders arranged at right angles to one another. They are important in cell division.
10. A CENTROSOME is a dense area of specialized cytoplasm located outside the nucleus that contains the centrioles.
11. FLAGELLA and CILIA have the same basic structure and are used in movement. If the projections are few, single, paired, or long they are called flagella. If they are numerous and hair-like, they are called cilia. Flagella can move an entire cell. Cilia move objects along a cell surface.

J. CYTOSKELETON

1. The CYTOSKELETON includes microfilaments, microtubules, and intermediate filaments.
2. MICROFILAMENTS are rod-like structures involved in the contraction of muscle cells. In other cells they provide support, shape, and assist in the locomotion of entire cells.
3. MICROTUBULES are slender tubes and provide support and shape for cells. In the cytosol they serve as channels for the movement of materials. They also provide the framework for cilia, flagella, and the mitotic spindle.
4. INTERMEDIATE FILAMENTS provide structural reinforcement in some cells and assist in contraction in others.

K. FLAGELLA AND CILIA

1. These are projections that some cells use for the movement of materials.
2. The only example of a cellular structure with a FLAGELLUM in the human is the spermatozoa.
3. CILIA are short and numerous projections, resembling hairs, found in the respiratory and reproductive tracts.

L. CENTROSOME AND CENTRIOLES

1. A CENTROSOME is a dense area of cytosol near the nucleus of the cell. It is involved in reproduction.
2. The centrosome has a pair of cylindrical structures, called CENTRIOLES, composed of microtubules and they play a role in the formation and regeneration of cilia and flagella.

M. PROTEIN SYNTHESIS

1. Cells make proteins by translating the genetic information encoded in the DNA into specific proteins. This involves TRANSCRIPTION and TRANSLATION.
2. In TRANSCRIPTION, genetic information encoded in the DNA is passed to a strand of messenger RNA (mRNA).
3. DNA also synthesizes ribosomal RNA (rRNA) and transfer RNA (tRNA).
4. In TRANSLATION, the information in the nitrogenous base sequence of the mRNA dictates the amino acid sequence of a protein.
5. PROTEIN SYNTHESIS occurs on ribosomes where a strand of mRNA is attached.
6. tRNA transports specific amino acids to the mRNA at the ribosome. A portion of the tRNA has a triplet of bases called an anticodon. Its complementary bases match the codon of three bases on the mRNA molecule.
7. The RIBOSOMES move along the mRNA strand as amino acids are joined to form the protein molecule.

N. NORMAL CELL DIVISION

1. MITOSIS (a nuclear division) and CYTOKINESIS (division of the cytoplasm and organelles) result in the production of somatic or gametic cells.
2. MEIOSIS (to be covered in Chapter 23) followed by cytokinesis, results in the production of the sex cells, either sperm or egg cells.
3. Prior to mitosis and cytokinesis, the DNA molecules, or chromosomes, replicate themselves so that the same hereditary traits will be passed on to future generations of cells.

O. SOMATIC CELL DIVISION

1. Human cells contain 23 pairs of chromosomes for a total of 46.
2. A CHROMOSOME is a DNA molegameticcule that stores hereditary information in genes.
3. DNA replication occurs in INTERPHASE. Also during this phase RNA and proteins are reproduced. The DNA appears as a coiled mass called CHROMATIN.

P. MITOSIS

1. Mitosis is the distribution of two sets of chromosomes into separate and equal nuclei following their replication. It consists of prophase, metaphase, anaphase, and telophase.
2. DNA replication occurs during INTERPHASE.
3. In PROPHASE, the chromosomes condense, the nuclear membrane and nucleoli disappear, centrioles migrate to opposite poles, and the chromosomes become visible.
4. In METAPHASE, the spindle fibers appear and the chromosomes align vertically along the equator of the cell.

5. In ANAPHASE, the centromeres divide and the single-stranded chromosomes move along the spindle fibers to the opposite poles.
6. In TELOPHASE, the chromosomes stretch out and unwind, a new nucleus and nucleoli appear, and regrouping is complete.
7. Finally, there is a division of the cytoplasm as cytokinesis occurs.
8. Cytokinesis usually begins in late anaphase and terminates in telophase.
9. A cleavage furrow forms at the cell's equatorial plane and progresses inward, cutting through the cell to form two separate portions of the cytoplasm.

Q. ABNORMAL CELL DIVISION: CANCER

1. Cancerous tumors (NEOPLASMS) can be referred to as being malignant or benign.
2. The study of cancer is called ONCOLOGY.
3. The spread of cancer from its primary site is called metastasis.
4. Carcinogens include environmental agents and viruses.
5. Treating cancer is difficult because all the cells in a single population do not behave the same way.

IV. TEACHING TIPS AND SUGGESTIONS

A. HELPFUL HINTS

1. Examine models, slides, transparencies, and photographs of a cell and the various organelles.
2. In the laboratory, prepared slides of various types of cells should be observed.
3. Demonstrate osmosis by filling a dialysis bag with 1% colored glucose solution. Tie both ends and place it in a beaker of 10% glucose solution. Have the students periodically check and observe the results.
4. Have students observe models in the various stages of mitosis.

B. ESSAY QUESTIONS

1. The systems of the body are capable of performing specialized functions that keep you alive and enable you to reproduce. In a very general kind of way, a cell is also capable of performing specific functions that maintain it on a daily basis and enable it to reproduce. Itemize the various structures in the cell and their contribution to the overall functioning of the cell.
2. You have just taken a sample of blood from a patient and wish to store it for several hours before running some chemical tests. Describe the composition of an isotonic solution in which you would place the cells. Explain the consequences if the red blood cells were to be placed in a hypertonic solution and why. What would be the effect if they were placed in a hypotonic solution?

C. Topics for Discussion

1. Do you feel that the aging process can be slowed down or reversed?
2. Compare the similarities between the mitochondrion and that of a simple bacterium. Is there any probability that the mitochondrion may once have been a completely independent organism?
3. Discuss how an adult can lose billions of body cells each day and still survive.
4. Manipulating the sequences of nitrogenous bases on the DNA double helix could have what possible consequences? Would such a process be detrimental, beneficial, or both?

V. AUDIOVISUAL MATERIALS

A. Overhead Transparencies

1. PAP Transparency Set, 8/e (Trs. 3.1-3.4, 3.6, 3.7a&b, 3.9, 3.10, 3.12, 3.17, 3.20, 3.21- 3.25f-h, 3.27a&b).
2. Cell Division Set (10 Transparencies).
3. Cell Machinery Set (12 Transparencies; CARO).
4. Cells and their Organelles (10 Transparencies; CARO).
5. Membranes (20 Transparencies; BM).

B. Videocassettes

1. Inside the Cell: Mechanisms and Molecules (CFH).
2. Exploring the Cell: Structure and Organelles (CFH).
3. Mitosis and Meiosis: How Cells Divide (CFH).
4. Cell Structure and Function (EIL).
5. Membranes (EIL).
6. Protein Synthesis (EIL).
7. DNA: Master Molecule of Life (1987; EIL).
8. The Genetic Gamble (Nova; 58 min.; 1985; COR/KSU).
9. Genetic Engineering (EIL).
10. The Body Ages (25 min.; 1987; KSU).
11. Cancer: The Causes (29 min.; 1983; KSU).
12. Toxic Trials (Nova: Leukemia; 58 min.; 1986; KSU).
13. The Living Cell (15 min.; C; Sd; 1990; IM).
14. The Cell (29 min.; C; Sd; 1990; IM).
15. Aging (26 min.; C; Sd; 1990; FHS).

C. Films: 16 mm

1. The Cell: A Functioning Structure, Parts I/II (30 min. ea.; 1972; CRM/McG/KSU).
2. The Living Cell (27 min.; 1972; H&R).
3. The Cell: Structural Unit of Life (11 min.; COR).
4. The Living Cell: DNA (20 min.; 1974; EBEC).
5. Mitosis (24 min.; 1980; EBEC).
6. Macromolecular Synthesis, Parts I-IV (KCI).

7. Inheritance in Man (28 min.; 1961; McG).
8. Can't It Be Anyone Else? (30 min.; 1980; PYR/KSU).

D. TRANSPARENCIES: 35 MM (2 x 2)

1. PAP Slide Set (Slides 10-16).
2. Visual Approach to Histology: Cytology (20 Slides; K&E).
3. DNA Basic Structure (K&E).
4. The Cell Cycle (63 Slides; BM).
5. Meiosis (74 Slides; BM).
6. Membranes (80 Slides; EIL).
7. Cell Motility (75 Slides; BM).
8. Protein Synthesis (62 Slides; EIL).
9. DNA (20 Slides; EIL).
10. Recombinant DNA (20 Slides; EIL).
11. Genetic Engineering (63 Slides; BM).
12. Cancer, The Disease (79 Slides; EIL).

E. COMPUTER SOFTWARE

1. Transcription/Translation (Apple IIe; Eastern Kentucky University).
2. DNA, Level I/II (Apple IIe; EME).
3. Gene Structure and Function (Apple IIe; EduTech).
4. Meiosis Level I/II (Apple IIe; EME).
5. Cell Structure and Function (Apple; IBM; EIL).
6. Gene Machine (Apple; EIL).
7. Mitosis (Apple; IBM; EIL).
8. Enzymes (Apple; EIL).

CHAPTER 4

TISSUES

CHAPTER AT A GLANCE

- TYPES OF TISSUE
- EPITHELIAL TISSUES
- *General Features of Epithelial Tissues*
- *Covering and Lining Epithelium*
- *Classification by Cell Shape*
- *Classification by Arrangement of Layers*
- *Types and Functions*
- *Simple Epithelium*
- *Stratified Epithelium*
- *Pseudostratified Epithelium*
- *Glandular Epithelium*
- CONNECTIVE TISSUE
- *General Features of Connective Tissue*
- *Connective Tissue Cells*
- *Connective Tissue Matrix*
- *Classification of Connective Tissue*
- *Embryonic Connective Tissue*
- *Mature Connective Tissue*
- *Loose Connective Tissue*
- *Dense Connective Tissue*
- *Cartilage*
- *Bone (Osseous) Tissue*
- *Blood (Vascular) Tissue*
- MEMBRANES
- *Mucous Membranes*
- *Serous Membranes*
- *Cutaneous Membranes*
- *Synovial Membranes*
- MUSCLE TISSUE
- NERVOUS TISSUE
- WELLNESS FOCUS: FAT TISSUE ISSUES

I. CHAPTER SYNOPSIS

The principal concern of this chapter is to examine the organization of cells into tissues. The structure, function, and location of the principal kinds of tissues are examined. Attention is given to the four major tissue types and their classification. Throughout, the relationship of structure to function is emphasized. Both embryonic and adult tissues are discussed. The chapter concludes with a discussion of the four different types of membranes.

II. LEARNING GOALS/STUDENT OBJECTIVES

1. Describe the general features of epithelial tissue.
2. Explain how covering and lining epithelium are classified.
3. Define the various types and functions of covering and lining epithelium.
4. Define a gland and distinguish between exocrine and endocrine glands.
5. Explain how connective tissue is classified.
6. Describe the various types of membranes in the body.

III. SAMPLE LECTURE OUTLINE

A. Types of Tissue

1. A TISSUE is a group of similar cells and their intracellular substances that function together to perform a specialized activity.
2. HISTOLOGY is the study of tissues.
3. The major tissue types are EPITHELIAL, CONNECTIVE, MUSCULAR, and NERVOUS.

B. Epithelial Tissue

1. EPITHELIAL tissue is the only tissue that is constantly regenerated in large amounts.
2. The two types of epithelium are LINING EPITHELIUM and GLANDULAR EPITHELIUM. Epithelial tissue contains very little intracellular material and is avascular.

C. General Feature of Epithelial Tissue

1. The numerous features of epithelial tissue are:
   - Epithelium consists of closely packed cells with little extracellular material.
   - Epithelial cells have free surfaces, exposed to a body cavity, lining of an organ, or exterior of body.
   - Epithelial tissues are avascular.
   - Epithelial tissues do not have a nerve supply.
   - Epithelial tissues have a high capacity for renewal.
   - Epithelial tissues function in protection, secretion, absorption, excretion, sensory reception and reproduction.

D. COVERING AND LINING EPITHELIUM

1. Epithelial tissue can be classified by SHAPE being squamous, cuboidal, columnar or transitional.
2. Epithelium can also be classified by the ARRANGEMENT OF LAYERS, being simple (single-layered), stratified (multi-layered), or pseudostratified (a layer of mixed cell shapes). Certain cells can also contain cilia and/or secrete mucus.
3. SIMPLE SQUAMOUS EPITHELIUM functions in the diffusion of gases and the movement of fluid and dissolved substances by osmosis.
4. SIMPLE CUBOIDAL EPITHELIUM functions to secrete tears and saliva and absorb water in the kidney tubules.
5. SIMPLE COLUMNAR EPITHELIUM may have modifications such as microvilli, which serve to increase the surface area or goblet cells which secrete mucus.
6. STRATIFIED SQUAMOUS EPITHELIUM usually contains a protein called keratin and serves as a protective barrier against microbes and water.
7. STRATIFIED CUBOIDAL EPITHELIUM is rare but functions in protection.
8. STRATIFIED COLUMNAR EPITHELIUM functions in protection and secretion.
9. TRANSITIONAL EPITHELIUM ALLOWS for organ expansion and prevents organ rupture.
10. PSEUDOSTRATIFIED COLUMNAR EPITHELIUM is a functional tissue composed of ducts and tubes throughout the entire body and provides for the movement of materials.
11. A SUMMARY OF EPITHELIAL TISSUES:

TYPES OF TISSUE	LOCATION	FUNCTION
SIMPLE SQUAMOUS	LINES HEART, LYMPHATIC VESSELS AND ABDOMINAL CAVITY, AND SEROUS SECRETIONS	FILTRATION, ABSORPTION
SIMPLE CUBOIDAL	COVERS OVARIES, LINES KIDNEY TUBULES AND PART OF THE EYE LENS, AND FORMS PART OF THE RETINA	SECRETION AND ABSORPTION
SIMPLE COLUMNAR	LINES GI TRACT FROM STOMACH EXCRETORY DUCTS OR GLANDS, AND GALL BLADDER	SECRETION AND ABSORPTION
SIMPLE CILIATED COLUMNAR	LINES UPPER RESPIRATORY TRACT, UTERINE TUBES, AND UTERUS	MOVES MUCOUS BY CILIARY ACTION
STRATIFIED SQUAMOUS	LINES THE MOUTH, TONGUE, ESOPHAGUS, VAGINA, AND OUTER LAYER OF THE SKIN	PROTECTION
STRATIFIED CUBOIDAL	COMPRISES DUCTS OF SWEAT GLANDS, CONJUNCTIVA OF THE EYE, PHARYNX, AND MALE URETHRA	PROTECTION

TYPES OF TISSUE	LOCATION	FUNCTION
STRATIFIED COLUMNAR	LINES MALE URETHRA	PROTECTION AND SECRETION
TRANSITIONAL	LINES URINARY BLADDER AND PARTS OF URETER AND URETHRA	PREVENTS DISTENTION
PSEUDOSTRATIFIED COLUMNAR	LINES LARGE EXCRETORY DUCTS, EPIDIDYMIS, MALE URETHRA, AND EUSTACHIAN TUBES, UPPER SPERM RESPIRATORY TRACT AND MALE REPRODUCTIVE SYSTEM	SECRETION AND MOVEMENT

E. GLANDULAR EPITHELIUM

1. The principal function of glandular epithelium is SECRETION.
2. A gland consists of one cell or a group of highly specialized epithelial cells secreting substances into ducts, onto a surface, or into the blood.
3. EXOCRINE GLANDS secrete products into ducts which empty onto an external or internal surface. Examples are sweat glands, salivary glands, and digestive tract glands.
4. ENDOCRINE GLANDS are ductless, release hormones, and secrete their products directly into circulation. Some examples are the pituitary and adrenal glands.

F. CONNECTIVE TISSUE

1. CONNECTIVE TISSUE is the most abundant tissue in the body and is highly vascular. It also contains a varied number of cells with a large quantity of intracellular matrix.
2. The general functions of connective tissue are protection, support, binding, and storage.

G. GENERAL FEATURES OF CONNECTIVE TISSUE

1. The following are general features of connective tissue:
   - Consists of three basic elements: cells, ground substance, and fibers.
   - The three basic components form the matrix.
   - Connective tissues do not generally occur on free surfaces.
   - Connective tissue has a nerve supply, except for cartilage which does not.
   - Connective tissue is highly vascular, except for cartilage which is avascular.
   - The matrix can be gelatinous, fibrous, or calcified.

H. CONNECTIVE TISSUE CELLS

1. Various types of cells are found in connective tissue. These include:
   - Fibroblasts
   - Macrophages
   - Plasma Cells
   - Mast Cells

I. CONNECTIVE TISSUE MATRIX

1. Each type of connective tissue has unique properties due to its matrix between the cells. The matrix contains proteins fibers embedded in the fluid, gel, or solid GROUND SUBSTANCE. The ground substance also contains several other substances. These include HYALURONIC ACID and CHONDROITIN SULFATE.
2. HYALURONIC ACID functions to bind cells together, lubricates joints, and helps maintains the shape of the eyeballs.
3. CHONDROITIN SULFATE provides support and adhesiveness in cartilage, bone, the skin, and blood vessels.
4. The ground substance also contains FIBERS which provide strength and support for the tissue. The major type of fibers include:
   - COLLAGEN which is very tough and resistant to pulling forces. These fibers occur in bundles and afford great strength. Collagen is the most abundant protein in the human body.
   - ELASTIC FIBERS are usually smaller than collagen. They are comprised of the protein ELASTIN which provides both strength and elasticity.
   - RETICULAR FIBERS are comprised of collagen and glycoprotein, and provide strength and support in the walls of blood vessels.

J. CLASSIFICATION OF CONNECTIVE TISSUES

1. The two types of connective tissue are EMBRYONIC CONNECTIVE TISSUE, which includes MESENCHYME and MUCOUS CONNECTIVE TISSUE, and ADULT CONNECTIVE TISSUE, which includes LOOSE, ADIPOSE, DENSE, ELASTIC and RETICULAR.
2. ADULT tissues also include the cartilages - hyaline, elastic, and fibrous.
3. OSSEOUS TISSUE (bone) and VASCULAR TISSUE (blood) are specialized connective tissues.
4. A SUMMARY OF EMBRYONIC CONNECTIVE TISSUES:

TYPE	DESCRIPTION	LOCATION	FUNCTION
mesenchyme	mesenchymal cells in a fluid matrix	embryo	forms all connective tissues

5. A SUMMARY OF MATURE CONNECTIVE TISSUES:

TYPE	DESCRIPTION	LOCATION	FUNCTION
areolar	fibers of collagen, elastic and reticular macrophages, plasma cells in a matrix	around organs, dermis of skin and tissue with fibroblasts, subcutaneous layer	strength, elasticity, and support
adipose	signet cells used for fat storage	subcutaneous tissue around heart and kidneys, marrow of long bones, joints	heart loss reduction, support, protection, and energy reserve
collagenous	collagen fibers with fibroblasts	tendons, ligaments protection, heart, bone, liver, testes and lymph nodes	support and attachment
elastic	elastic fibers with few fibroblasts	lungs, arteries, trachea, bronchi, and true vocal cords	stretching
reticular	reticular fibers with cells wrapped around	liver, spleen, lymph nodes	forms stroma and basal laminae of organs
hyaline cartilage	gristle; glossy mass with chondrocytes	end of long bones, parts of the larynx, bronchi and embryonic skeleton	movements at joints, support and flexibility
fibrocartilage	chondrocytes in bundles of collagenous fibers	joints between hip bones, intervertebral discs and knees	support and fusion
elastic cartilage	chondrocytes in a network of elastic fibers	laryngeal epiglottis, external ear and Eustachian tubes	support and shape

6. Connective tissue may contain a variety of CELL TYPES such as fibroblasts, macrophages, plasma cells, mast cells, melanocytes, adipocytes, and leukocytes.
7. OSSEOUS TISSUE is a specialized connective tissue comprised of bone-forming cells (osteocytes), and can exist either as compact or cancelleous bone.
8. The OSTEON or HAVERSIAN SYSTEM is the basic unit of compact bone consisting of lamellae (matrix system), lacunae (small spaces between the lamellae with osteocytes), canaliculi (minute canals projecting from lacunae), and the Haversian canal (containing the principal blood vessel).
9. VASCULAR TISSUE is a specialized liquid connective tissue of plasma and formed elements. The formed elements are erythrocytes, leukocytes, and thrombocytes.

K. MUSCLE TISSUE

1. MUSCLE TISSUE is a highly specialized tissue for contraction, motion, the maintenance of posture, and the production of heat.
2. SKELETAL MUSCLE is attached to bones, is voluntary, striated multi-nucleate and allows for movement. The plasma membrane is referred to as the sarcolemma and the cytoplasm is referred to as the sarcoplasm.
3. CARDIAC MUSCLE forms the bulk of the heart and its contraction allows for the pumping of blood. It is involuntary and striated. The cardiac muscle cells, which are uninucleate, are separated from one another by thickenings of the sarcolemma called intercalated discs.
4. SMOOTH MUSCLE is found in most of the body organs and is also referred to as visceral muscle. It is involuntary, uninucleate, non-striated, and has spindle-shaped cells.

L. NERVOUS TISSUE

1. NERVOUS TISSUE contains two principal tissue types: NEURONS and NEUROGLIA.
2. NEURONS respond to stimuli, conduct impulses to the brain and spinal cord, other neurons, muscle fibers, and glands.
3. Neurons are comprised of a cell body with organelles, dendrites (specialized structures which convey impulses to the cell body), and axons (structures that conduct impulses from the cell body).
4. NEUROGLIA are the cellular support for nervous tissue.

M. EPITHELIAL MEMBRANES

1. EPITHELIAL MEMBRANES are a combination of an epithelial layer and underlying connective tissue. Three types of epithelial membranes are MUCOUS, SEROUS, and CUTANEOUS. SYNOVIAL MEMBRANES are specialized epithelial membranes which do not contain any epithelial cells.

IV. TEACHING TIPS AND SUGGESTIONS

A. HELPFUL HINTS

1. A lecture on tissues should be accompanied by either overhead photomicrographs or 35 mm transparencies so that the student can see the similarities and differences between tissues.
2. Laboratory microscopic examination should consist of viewing prepared tissue slides accompanied by 35 mm slides and photomicrographs.

3. A model of the osteon will aid in the understanding of compact bone structure. Bones from the skeleton, exhibiting cancelleous and compact bone, will serve as an excellent aid.

B. Essay Questions

1. Distinguish covering and lining epithelium from glandular epithelium. What characteristics are common to all epithelial tissue?
2. List the major ways in which connective tissues differ from epithelial tissues.
3. How are embryonic connective tissue and mature connective tissue distinguished?
4. Distinguish between neurons and neuroglia.

V. AUDIOVISUAL MATERIALS

A. Overhead Transparencies

1. PAP Transparency Set (Trs. 4.1a-e, 4.2 and 4.3).
2. Electron Micrographs of Muscle Set (11 Transparencies; CARO).
3. Electron Micrographs of Nervous Tissue Set (11 Transparencies; CARO).
4. Electron Micrographs of Blood (10 Transparencies; CARO).
5. Electron Micrographs of Vascular Tissue (6 Transparencies; CARO).
6. Human Bone Histology (CARO).
7. Human Skin Histology (CARO).

B. Videocassettes

1. Accident (26 min.; S; Sd; 1989; FHS).
2. Cytology and Histology (25 min.; C; Sd; 1990; IM).
3. Tissues (29 min.; C; Sd; 1990; IM).

C. Films: 16 mm

1. Tissues of the Human Body (16 min.; C; Sd; 1963; KSU).
2. From One Cell (14 min.; C; Sd; 1990; IM).

D. Slides: 35 mm (2 x 2)

1. PAP Histology Set (Slides 1-157; specifically 1-58).
2. Animal Cells and Tissues I, II (160 Slides; CARO).
3. Histology (100 Slides; EI).
4. Human Tissues (NSC).
5. Tissues and Organ Set (262 Slides; REX).

E. Computer Software

1. Flash: Medical Terms (Apple; IBM; PLP).
2. Health Awareness Games (Apple II; IBM PC/jr.; Queue).
3. Health Risk Appraisal (Apple IIs; IBM PC/jr.; Queue).

CHAPTER 5

INTEGUMENTARY SYSTEM

CHAPTER AT A GLANCE

- SKIN
- *Structure*
- *Functions*
- *Epidermis*
- *Dermis*
- *Skin Color*
- ACCESSORY ORGANS OF THE SKIN
- *Hair*
- *Glands*
- *Sebaceous (Oil) Glands*
- *Sudoriferous (Sweat) Glands*
- *Ceruminous Glands*
- *Nails*
- AGING AND THE INTEGUMENTARY SYSTEM
- HOMEOSTASIS OF THE BODY TEMPERATURE
- COMMON DISORDERS
- MEDICAL TERMINOLOGY AND CONDITIONS
- WELLNESS FOCUS: MORE THAN SKIN DEEP

I. CHAPTER SYNOPSIS

Students are introduced to the organ and systems' levels of organization by considering the anatomy and physiology of the skin and its epidermal derivatives, such as hair, nails, and various glands. Factors contributing to skin color and wound healing are discussed. The role of the integumentary system in the homeostasis of body temperature as part of a negative feedback system is developed. Consideration is given to some common disorders of the skin such as burns, acne, decubitus ulcers, sunburn, and cancer. The chapter concludes with a list of medical terms and conditions associated with the integumentary system.

II. LEARNING GOALS/STUDENT OBJECTIVES

1. Describe the structure and functions of the skin.
2. Explain the pigments involved in skin color.
3. Describe the structure and functions of the accessory organs of the skin.
4. Describe the effects of aging on the integumentary system.
5. Explain how the skin helps regulate body temperature.

III. SAMPLE LECTURE OUTLINE

A. THE SKIN- STRUCTURE

1. The skin is an organ which consists of several types of tissues and performs several specialized activities.
2. The tissues comprising the skin are the epithelium of the EPIDERMIS and the connective tissues of the DERMIS.
3. The two principal layers of the skin are the epidermis and the dermis.

B. THE FUNCTIONS OF THE SKIN

1. The skin exhibits numerous functions. These include:
   • Regulation of body temperature
   • Protection
   • Sensation
   • Excretion
   • Immunity
   • Synthesis of Vitamin D

C. THE EPIDERMIS

1. The epidermis is composed of stratified squamous epithelium and contains several cell types.
2. The CELL TYPES found in the epidermis are KERATINOCYTES, MELANOCYTES, LANGERHANS' CELLS, and GRANSTEIN CELLS.
3. The epidermis is organized into five DISTINCT LAYERS. These are the STRATUM CORNEUM, STRATUM SPINOSUM, STRATUM GRANULOSUM, STRATUM LUCIDUM, and STRATUM BASALE.
4. The skin is capable of regeneration.

D. THE DERMIS

1. The DERMIS consists of two distinct regions. The upper, PAPILLARY REGION, is composed of loose, connective tissue and elastic fibers; and the lower, RETICULAR LAYER, is composed of dense, irregularly-arranged connective tissue and collagenous and elastic fibers.
2. Lines of cleavage on the surface of the skin overlay dermal regions where collagen fibers are oriented in a specific fashion. These are referred to as the DERMATOMES.
3. SKIN COLOR is due to MELANIN in the epidermis, CAROTENE in the dermis, and blood in the capillaries of the dermis. Differences in skin color are due to the amount of melanin produced and the extent of its dispersal.
4. MALIGNANT MELANOMA, (cancer of the melanocytes), is a particularly serious skin cancer. Liver, or age spots, are non-cancerous clusters of melanin.

E. ACCESSORY ORGANS OF THE SKIN

1. The accessory organs of the skin include HAIR, GLANDS, and NAILS.
2. HAIR is distributed variously over the body where it functions in protection. Each hair is composed of a SHAFT and a ROOT. Surrounding the root is the HAIR FOLLICLE, which is composed of two layers of epidermal tissue.

F. GLANDS

1. Accessory glands of the skin include SEBACEOUS (OIL), SUDORIFEROUS (SWEAT), and CERUMINOUS (WAX) GLANDS.
2. SEBACEOUS GLANDS, with few exceptions, are associated with the hair follicle and secrete an oily substance called SEBUM which prevents skin dryness and water evaporation.
3. Blackhead or pimples represent enlargement of the sebaceous glands due to unreleased quantities of sebum.
4. SUDORIFEROUS GLANDS, or SWEAT GLANDS, can be divided into APOCRINE GLANDS and ECCRINE GLANDS.
5. APOCRINE GLANDS are found in the armpit, pubic region, and the pigmented area of the breasts.
6. ECCRINE GLANDS are found throughout the skin except the lips, nail beds, glans penis, glans clitoris, labia minora, and eardrums.
7. CERUMINOUS glands are modified sudoriferous glands present in the external ear. They produce a sticky substance called CERUMEN which provides a barrier against foreign bodies.

G. NAILS

1. NAILS are hard keratinized epidermal cells over the dorsal surfaces of the terminal portions of the fingers and toes.

H. HOMEOSTASIS AND BODY TEMPERATURE

1. The skin helps to regulate the homeostasis of body temperature through a negative feedback system.
2. Perspiration from sweat glands and dilation of the superficial blood vessels help to remove excess heat from the body.
3. Constriction of the blood vessels in the skin aids in conserving heat when body temperature drops.
4. BURNS can destroy the proteins in the exposed cells and cause injury or death. Burns are classified by depth.
5. FIRST-DEGREE and SECOND-DEGREE burns are called partial-thickness burns.
6. THIRD-DEGREE burns are termed full-thickness burns.
7. The LUND-BROWDER method is one which is used to approximate the extent of burns. The rule of nines can be used for a more rapid, if somewhat less accurate, assessment of the extent of burns.

IV. TEACHING TIPS AND SUGGESTIONS

A. HELPFUL HINTS

1. Most students have questions concerning skin products and conditions. You may want to ask each student to write down a questions that the class can discuss the involvement of the integumentary system with skin products and conditions.
2. Stress the importance of the terminology associated with the integumentary system.

B. Essay Questions

1. Describe the activities of the skin in the regulation of a constant body temperature when the environmental temperature is about 100 degrees Fahrenheit. Why are these activities considered a negative feedback system? What happens to the skin when the temperature goes below freezing?
2. What are "pores" of the skin and what secretion is discharged through these structures? What effect, if any, would you expect acne medication creams to have on these pores?
3. List three different ways that the integument is affected by aging. Suggest some ways by which one may minimize the aging process.

C. Topics for Discussion

1. Discuss the use of collagen creams. Are they effective and can they be absorbed through the skin?
2. Discuss what effect the presence of sunlight has on the pigmentation of the skin. Would you expect to find differences in pigmentation between individuals living in the dark as opposed to those that were exposed to sunlight? What would be the major effects of not being exposed to sunlight?

V. AUDIOVISUAL MATERIALS

A. Overhead Transparencies

1. PAP Transparency Set (Trs. 5.1, 5.2a&b, 5.3a, 5.6a&b & 5.7).
2. IM Transparency Masters (TM 1).

B. Videocassettes

1. The Skin: Its Structure and Function (21 min.; 1983; EBEC/KSU).
2. Hot and Cold (26 min.; FHS).

C. Films: 16 mm.

1. The Skin: Its Structure and Function (21 min.; 1983; EBEC/KSU).
2. Regulation of Body Temperature (22 min.; EBEC).
3. Your Skin (15 min.; 1964; FAD/KSU).
4. Skin Deep (26 min.; FHS).
5. I Am Joe's Skin (30 min.; 1984; PYR/KSU).

D. Transparencies: 35 mm. (2x2)

1. PAP Slide Set (Slides 17-18).
2. AHA Slide Set.
3. Integumentary System (Slides 1-5; McG).
4. Visual Approach to Histology: Integumentary System (12 Slides; FAD).
5. Systems of the Human Body: Skin and Its Functions (20 Slides; EIL).

E. COMPUTER SOFTWARE

1. Dynamics of the Human Skin (Apple II Series; IBM PC; 1988; C3059; EIL).
2. Skin Probe (Apple; SC-175032; PLP).

CHAPTER 6

THE SKELETAL SYSTEM

CHAPTER AT A GLANCE

- FUNCTIONS
- TYPES OF BONES
- PARTS OF A BONE
- HISTOLOGY
- *Compact Bone Tissue*
- *Spongy Bone Tissue*
- OSSIFICATION: BONE FORMATION
- *Intramembranous Ossification*
- *Endochondral Ossification*
- HOMEOSTASIS
- *Bone Growth and Maintenance*
- *Bone and Mineral Homeostasis*
- *Aging and Bone*
- *Exercising and Bone*
- BONE SURFACE MARKINGS
- DIVISIONS OF THE SKELETAL SYSTEM
- SKULL
- *Sutures*
- *Cranial Bones*
- *Facial Bones*
- *Fontanels*
- *Foramina*
- HYOID
- VERTEBRAL COLUMN
- *Divisions*
- *Normal Curves*
- *Typical Vertebra*
- *Cervical Region*
- *Thoracic Region*
- *Lumbar Region*
- *Sacrum and Coccyx*
- THORAX
- *Sternum*
- *Ribs*
- PECTORAL (SHOULDER) GIRDLE
- *Clavicle*
- *Scapula*
- UPPER EXTREMITY
- *Humerus*
- *Ulna and Radius*

- *Carpals, Metacarpals, and Phalanges*
- PELVIC (HIP) GIRDLE
- LOWER EXTREMITY
- *Femur*
- *Tibia and Fibula*
- *Tarsals, Metatarsals, and Phalanges*
- *Arches of the Foot*
- COMPARISON OF MALE AND FEMALE SKELETONS
- COMMON DISORDERS
- MEDICAL TERMINOLOGY AND CONDITIONS
- WELLNESS FOCUS: MAKING AN IMPACT ON BONE DENSITY

I. CHAPTER SYNOPSIS

The chapter considers the major functions of the skeletal system, principal bone types, histology of bone, and both intramembranous and endochondral ossifications. Students are then introduced to important surface markings of bone, the divisions of the skeletal system, and the various bones of the axial and appendicular skeletons. A detailed description of all bones is then presented. The chapter concludes with differences between the male and female skeletons, common skeletal disorders, and a list of medical terms.

II. LEARNING GOALS/STUDENT OBJECTIVES

1. Discuss the functions of the skeletal system.
2. Describe the microscopic structure of compact and spongy bone tissue.
3. Explain the steps involved in bone formation.
4. Describe the factors involved in bone growth and maintenance.
5. Classify the bones of the body into axial and appendicular divisions.
6. Describe the structural and functional features of the vertebral column.
7. Identify the bones of the upper extremity.
8. Identify the bones of the lower extremity.
9. Compare the principal structural and functional differences between male and female skeletons.

III. SAMPLE LECTURE OUTLINE

A. SKELETAL SYSTEM FUNCTIONS

1. The skeletal system performs the following functions:
   - Support
   - Protection
   - Movement
   - Mineral storage and homeostasis
   - Site of blood cell production
   - Storage of energy

B. TYPES OF BONES

1. Bones can be classified on the basis of shape into four principal types. These are LONG, SHORT, FLAT, and IRREGULAR.
2. LONG BONES have a greater length than width and consist of a main central cylinder called the SHAFT, or DIAPHYSIS, and extremities, called the EPIPHYSIS. These are comprised mostly of COMPACT TISSUE.
3. SHORT BONES are somewhat cubed-shaped and nearly equal in length and width. They are mostly SPONGY BONE except at the surface.
4. FLAT BONES are thin and comprised of two or more or less parallel plates of compact tissue enclosing spongy bone.
5. IRREGULAR BONES have complex shapes and vary in composition of compact and spongy bone.

C. PARTS OF A LONG BONE

1. The typical anatomy of a long bone includes the following:
   - DIAPHYSIS- shaft or long cylinder of a bone
   - EPIPHYSIS- the extremities of a long bone
   - METAPHYSIS- the area where the diaphysis and epiphysis meet
   - ARTICULAR CARTILAGE- a thin layer of hyaline cartilage covering the epiphyseal ends at an articulation point with another bone
   - PERIOSTEUM- the tough, white fibrous connective tissue on the outer layer of the bone
   - MEDULLARY (MARROW) CAVITY- specifically within the diaphysis it contains yellow bone marrow in the adult
   - ENDOSTEUM- the lining of the medullary canal which consists of OSTEOPROGENITOR CELLS and OSTEOCLASTS

D. HISTOLOGY OF BONE

1. Structurally, several types of connective tissue are involved in the skeletal system. These include cartilage, dense connective tissue, and osseous tissue.
2. Bone is connective tissue with a great deal of intracellular substance (MATRIX) and sparsely distributed cells. The intracellular substance consists of calcium, phosphate salts, and collagenous fibers. The collagenous fibers form the framework around which the salts will crystallize.
3. The types of cells found in bone tissue are OSTEOPROGENITOR (OSTEOGENIC) cells, OSTEOBLASTS, OSTEOCYTES, and OSTEOCLASTS.
4. The typical large bone has several parts. The microscopic structure is easier to understand if one is first familiar with its gross structure.
5. Bone is not solid. It contains a SOLID REGION (COMPACT OR DENSE), and a more POROUS REGION (CANCELLOUS OR SPONGY).
6. The structural unit of compact bone is the OSTEON or HAVERSIAN SYSTEM. It is characterized by a concentric ring structure or layer (LAMELLAE), a central (HAVERSIAN) CANAL which is surrounded by interspersed layers of matrix, and osteocytes in small spaces (LACUNAE).
7. The structural unit of cancellous bone is the TRABECULA, an irregular lattice-work of osseous tissue. Contained in the trabeculae are osteocytes and red bone marrow.

E. Compact Bone Tissue

1. Compact bone tissue contains free spaces and forms the external layer of all bones.
2. Compact bone tissue provides protection and support and aids in the stress of weight placed upon it.
3. Nutrient vessels penetrate compact bone perforating Volkman's canals and are connected to those of the central Haversian Canals.
4. The central canals run lengthwise and are surrounded by concentric lamellae.
5. The lamellae contain small spaces called lacunae which contain the osteocytes.
6. Radiating from the lacunae are the canaliculi which contain the slender process of the ostocytes. These interconnect all of the lacunae to the central canal.
7. The sum of the central canal, surrounding lamellae, lacunae, and osteocytes comprise the Haversian system or Osteon.

F. Spongy Bone Tissue

1. Spongy bone tissue does not contain true osteons. It is composed of an irregular latticework of thin plates of bone called trabeculae in which the microscopic spaces are filled with red bone marrow.
2. Spongy bone tissue comprises the bulk of all bones in the body and the spongy bone tissue of the pelvis, ribs, sternum, vertebrae, skull, and some of the epiphyseal ends, is the site of red bone marrow and hemopoiesis in the adult.

G. Ossification: Bone Formation

1. Bone formation, or ossification, begins with osteoprogenitor cells. These arise from mesenchymal embryonic connective tissue.
2. The osteoprogenitor cells give rise to chondroblasts which produce cartilage and osteoblasts, which form bone.
3. Bone formation takes place in the embryo, on or within fibrous membranes (intramembranous ossification), or in cartilage (endochondral ossification).
4. The surface skull bones and clavicles are formed intramembranously.
5. The other bones, including the lower skull bones, are formed by endochondral ossification.

H. Bone Growth and Maintenance: Homeostasis

1. The cartilage that exists between the epiphysis and diaphysis is the epiphyseal plate. It permits growth in length until adulthood.
2. Bone growth is under the control of the Human Growth Hormone (hGH) from the anterior pituitary gland as well as ovarian and testicular sex hormones.
3. Vitamin D is also necessary for the absorption of calcium ions from the digestive tract into the blood, proper bone mineralization, calcium removal from the bones, and reabsorption of calcium from potential urine via kidney tubules.
4. The ossification process is usually complete by age 25, and somewhat earlier in females.
5. When the ossification process is complete, the cartilage in the epiphyseal plate ossifies and forms the epiphyseal line.

I. Bone and Mineral Homeostasis

1. Bones store and release calcium and phosphate.
2. This is controlled by CALCITONIN (CT), PARATHYROID HORMONES (PTH), and VITAMIN D.
3. PTH activates the ostoclasts and decreases bone density by increasing serum calcium levels.
4. CT decreases ostoclastic activity and increases osteoblastic activity. It decreases serum and blood calcium levels, and increases in bone density.

J. Exercise and Bones

1. When placed under mechanical stress, bone tissue becomes stronger.
2. The important mechanical stresses result from the pull of skeletal muscles and the pull of gravity.

K. Aging and Bone

1. Aging is a result of calcium loss from the body.
2. It begins after the age of 30 in the female and accelerates between 40-45, as the estrogen levels decrease.
3. In the male, calcium loss does not generally begin until after the age of 60.

L. Bone Surface Markings

1. Specific surface markings which exist on bones generally indicate function; rough areas such as the deltoid TUBEROSITY, serve as attachment sites for muscle.
2. Bone markings are classified as DEPRESSIONS, OPENINGS, and PROCESSES that form joints to which tendons, ligaments, and other connective tissues attach.
3. The depressions amid openings are:
   - FORAMEN - opening for blood vessels, nerves, and ligaments
   - MEATUS - tube-like passageway
   - PARANASAL SINUS - air-filled cavity within bony connections to nasal cavity
   - FOSSA - depression in or on a bone
4. The processes that form joints are:
   - CONDYLE - large rounded prominence
   - HEAD - rounded projection supported by a constricted neck
   - FACET - smooth, flat surface
5. The process to which tendons, ligaments, and other connective tissue structures attach are:
   - TUBEROSITY - large, rounded, roughened process
   - SPINOUS PROCESS - sharp, slender process
   - TROCHANTER - large, blunt projection
   - CREST - prominent ridge or border

M. Sutures

1. Sutures are immovable joints of the body found only in the skull bones. They are represented by the following:
   - CORONAL SUTURE - unites the frontal bone with the two parietal bones
   - SAGITTAL SUTURE - unites the two parietal bones

• LAMBDOIDAL SUTURE - unites the occipital bones with the two parietal bones
• SQUAMOUSAL SUTURE - unites the two parietal bones with the temporal bone

N. AXIAL SKELETON: SKULL

1. The SKULL rests atop the vertebral column and consists of the CRANIAL BONES (8) and FACIAL BONES (14).
2. The CRANIAL BONES are the FRONTAL bone (1), PARIETAL bones (2), TEMPORAL bones (2), OCCIPITAL bone (1), SPHENOID bone (1), and ETHMOID bone (1).
3. The FACIAL bones are the NASAL bones (2), MAXILLA (2), ZYGOMATIC bones (2), MANDIBLE (1), LACRIMAL (2), PALATINE bones (2), INFERIOR NASAL CONCHAE (2), and the VOMER (1).

O. AXIAL SKELETON: HYOID BONE AND OSSICLES

1. The HYOID bone is unique in that it does not articulate with any other bone in the body. It acts as an attachment point for several muscles and ligaments of the tongue, neck, and pharynx.
2. The three OSSICLES are the INCUS, MALLEUS, and STAPES, also known as the hammer, anvil, and stirrup. They are formed in each tympanic cavity and are the smallest bones in the body.

P. AXIAL SKELETON: VERTEBRAL COLUMN

1. The VERTEBRAL COLUMN is composed of 26 bones, distributed into five regions.
2. The CERVICAL REGION, just below the skull contains 7 bones. The THORACIC REGION contains 12 bones. The LUMBAR region contains 5 bones. The SACRAL region contains 5 bones fused into one. The COCCYGEAL region can contain up to 4 bones fused into one.
3. There are FOUR CURVES found in the normal vertebral column. They aid to increase strength, maintain balance, absorb shocks, and prevent fracturing of the vertebral column bones.
4. The THORACIC and SACRAL CURVES are primary curves, remnants of the fetus's single anteriorly concave curve.
5. The CERVICAL and LUMBAR CURVES are anteriorly convex. These secondary curves develop as a child begins to hold the head up (cervical curve) and assumes an upright position (lumbar curve) respectively.
6. Between each vertebrae there is a disc of fibroelastic cartilage called the INTERVERTEBRAL DISC, which serves to act as a shock absorber.
7. The typical vertebra can be characterized by specific components. These are a BODY, PEDICLE, LAMINA, SPINOUS PROCESS, and TRANSVERSE PROCESS. Some exceptions do occur in the cervical and sacral regions.

Q. AXIAL SKELETON: THORAX

1. The THORACIC CAGE consists of the STERNUM, RIBS (and associated costal cartilages), and the BODIES OF THE THORACIC VERTEBRAE.
2. The STERNUM is composed of three major areas. These are the MANUBRIUM, GLADIOLUS and the XIPHOID process.
3. Twelve pairs of RIBS make up the thoracic cavity. They attach posterior to the thoracic vertebrae.
4. The FIRST SEVEN PAIRS of RIBS are attached directly to the sternum via hyaline cartilage and are called the TRUE RIBS.

3. Try to have your students identify the major markings on any bones with eyes closed or while holding the bone behind their back. This can be done with a lab partner.

B. Essay Questions

1. Draw one typical bone of each of the four principal types of bones. Draw surface markings and indicate their functions.
2. Why is the sternum frequently used as the site for a bone marrow biopsy?
3. Describe in detail the two arches of the bones of the foot, and name the key bones that help support the weight of the body on these arches.

C. Topics for Discussion

1. Discuss the role of the female pelvis, specifically the sacro-iliac joint, true and false pelvises, and symphysis pubis, in labor and delivery.
2. Explain the various skeletal adaptations present in the female but not present in the male. Why are these adaptations essential?

V. AUDIOVISUAL MATERIALS

A. Overhead Transparencies

1. PAP Transparency Set (Trs. 6.1a, 6.2, 6.3a-c, 6.4, 6.7, 6.8, 7.1a&b, 7.2-7.5, 7.6a&b, 7.7-7.9, 7.11a&b, 7.13a&b, 7.15a&b, 7.16a&b, 7.21, 7.22a&b, 8.1a&b, 8.6a&b, 8.8a&b, 8.9a&b, 8.13).
2. Skeletal System (HSC).
3. Skeletal and Muscle Set (6 Transparency Set; CARO).
4. Skeletal System Unit 2 (C, 27 Transparencies; RJB).
5. The Skeletal System (Parts 1 and 2; PLP).
6. Skeletal System (Transparencies 6-31, McG).

B. Videocassettes

1. Preventing Back Injuries (24 min.; KSU).
2. Oh, My Aching Back (28 min.; 1985; KSU).
3. Anatomy of the Upper Limbs (52 min.; CARO).
4. Locomotion and Skeletons (29 min.; C; Sd; 1978; CARO).

C. Films: 16 mm

1. The Spinal Column (11 min.; 1956; EBEC).
2. Skeletal System (12 min.; COR).
3. I Am Joe's Spine (25 min.; 1974; PYR/KSU).
4. A Technique of Total Hip Replacement (26 min.; LP).
5. The Human Body: Skeletal System (12 min.; 1980; COR/KSU).
6. The Skeleton (17 min.; 1979; EBEC/KSU).

D. TRANSPARENCIES: 35 MM (2x2)

1. PAP Slide Set (Slides 22-29).
2. AHA Slide Set.
3. Anatomy of the Skull, Parts I-IV (310 Slides; BM).
4. Anatomy of the Vertebral Column (870 Slides; BM).
5. Anatomy of the Upper Limbs (80 Slides; BM).
6. Anatomy of the Lower Limbs (80 Slides; BM).
7. X-Rays of the Human Body (20 Slides; EIL).

E. COMPUTER SOFTWARE

1. Body Language: Skeletal System I-II (Apple II Series; IBM-PC; ESP).
2. Bone Probe (Apple; SC-175027; PLP).
3. Skeletal System (Apple; SC-182005; IBM; SC-182006; PLP).
4. Flash: Bones and Joints (Apple; IBM; PLP).
5. Graphic Human Anatomy and Physiology Tutor: Skeletal System (IBM; PLP).
6. Support, Locomotion, and Behavior (Apple II; IBM-PC; Mac; Queue).

CHAPTER 7

ARTICULATIONS

CHAPTER AT A GLANCE

- CLASSIFICATION OF JOINTS
- *Functional*
- *Structural*
- SYNARTHROSIS (IMMOVABLE JOINTS)
- *Suture*
- *Gomphosis*
- *Synchrondrosis*
- AMPHIARTHROSIS (SLIGHTLY MOVABLE)
- *Syndesmosis*
- *Symphysis*
- DIARTHROSIS (FREELY MOVABLE)
- *Structure*
- *Types*
- *Gliding*
- *Hinge*
- *Pivot*
- *Condyloid*
- *Saddle*
- *Ball-and-Socket*
- *Special Movements*
- Knee Joint
- COMMON DISORDERS
- MEDICAL TERMINOLOGY AND CONDITIONS
- WELLNESS FOCUS: LIVING WITH ARTHRITIS

I. CHAPTER SYNOPSIS

This chapter introduces the student to the various types of joints in the body. Functionally, joints are classified as synarthroses, amphiarthroses, and diarthroses. On the basis of structure, joints are classified as fibrous, cartilaginous, and synovial. There is also an extensive discussion about the structure and types of movements afforded by synovial joints. Specific joints such as the humeroscapular, coxal, and tibiofemoral are discussed at length. The major joints of the body are then listed in terms of articulating components, classification, and movements. The chapter concludes with clinical applications, common arthrological disorders, and a summary of medical terminology of arthrology.

II. LEARNING GOALS/STUDENT OBJECTIVES

1. Define an articulation (joint) and describe how the structure of an articulation determines its function.
2. Describe the structure of a typical diarthosis.
3. Describe the types of diarthoses and the movements that occur with each.
4. Describe several special movements that occur at diarthoses.

III. SAMPLE LECTURE OUTLINE

A. Introduction

1. An ARTICULATION, or joint, is a point of contact between bones, between cartilage and bone, or between teeth and bones.
2. The structure of the joint determines how it will function.
3. The range of movements of a joint are IMMOVABLE, SLIGHTLY IMMOVABLE, or FREELY MOVABLE.
4. Articulations are supported by ligaments and joint capsules, as well as by the attachment of surrounding muscles.

B. Classification of Joints

1. The STRUCTURAL CLASSIFICATION of joints includes fibrous, cartilaginous, and synovial.
2. The FUNCTIONAL CLASSIFICATION of joints defines the DEGREE OF MOVEABILITY of the joint and includes SYNARTHROSES (immovable joints), AMPHIARTHROSES (slightly movable joints), and DIARTHROSES (freely movable joints).

C. Synarthrosis (Immovable Joints)

1. The three types of synarthrotic joints are SUTURES, SYNDESMOSES, and GOMPHOSES. They are held closely together by fibrous connective tissue and provide little or no movement.
2. SUTURES are found between the bones of the skull and are immovable. Functionally, they are classified as synarthrotic joints. Examples of sutures are the coronal suture, lambdoidal suture, and parietal suture.
3. SYNDESMOSES contain large amounts of fibrous connective tissue between the bones. They are slightly movable and are functionally classified as amphiarthrotic joints. An example of a syndesmosis is the distal articulation of the tibia and the fibula.
4. A GOMPHOSIS is a type of fibrous joint in which a peg fits into a socket such as the articulations between the teeth and the bones of the mandible and the maxilla. Gomphoses are always synarthrotic.

D. Amphiarthrosis (Slightly Movable Joint)

1. AMPHIARTHROTIC JOINTS are held together by cartilage and afford little or no movement. The two types of cartilaginous joints are SYNCHONDROSES and SYMPHYSES.
2. A SYNCHONDROSIS is a joint in which the binding material is hyaline cartilage. Synchondroses are functionally classified as synarthroses. Examples of synchondroses include the epiphyseal plate and the articulation of the first rib and the sternum.

3. A SYMPHYSIS is joined by a broad, flat, fibrocartilagenous disc. It affords slight movement and is classified functionally as an amphiarthrotic joint. Examples of a symphysis are the intervertebral discs and the symphysis pubis.

E. DIARTHROSIS (FREELY MOVABLE JOINTS)

1. The typical DIARTHROTIC JOINT contains a SYNOVIAL CAVITY lined with a SYNOVIAL MEMBRANE, an ARTICULAR CARTILAGE, and an ARTICULAR CAPSULE. Most synovial joints contain accessory ligaments, articular discs and bursae.
2. The ARTICULAR CAPSULE consists of an outer fibrous capsule and an inner synovial membrane. It encloses the synovial cavity and unites the articulating bones.
3. The SYNOVIAL MEMBRANE secretes synovial fluid which lubricates the joints and provides nourishment for the articular cartilage.
4. BURSAE are sac-like structures which are situated between bones and several other structures such as skin, tendons, ligaments, and muscles. Their function is to reduce friction between the moving parts.
5. The presence of a JOINT CAVITY permits movement. This movement is limited due to the shape of the articulating bones, tension of the ligaments and apposition of soft parts.
6. A synovial joint that permits movement along one plane is referred to as MONOAXIAL; in two planes, BIAXIAL; and in three planes TRIAXIAL. Joints that do not move about an axis are referred to as being NON-AXIAL.

F. TYPES OF DIARTHROSES

1. Diarthrotic joints are divided into six subtypes. These are GLIDING, HINGE, PIVOT, CONDYLOID, SADDLE, and BALL-AND-SOCKET.
2. Gliding joints contain an articular surface which is flat and allows a gliding movement. Examples are the joints between the carpals, tarsals, sternum and clavicle, and scapula and clavicle.
3. HINGE JOINTS allow one bone to move into the concave surface of another bone. These joints allow for flexion and extension as well as hyperextension and are exemplified by the elbow and knee joints.
4. PIVOT JOINTS contain a round or pointed surface of one bone which articulates within a ring formed by another bone. This allows for ROTATION and can be exemplified by the atlanto-axial joint.
5. CONDYLOID JOINTS contain oval-shaped bones which fit into a depression in another bone. This allows for ABDUCTION and ADDUCTION as well as CIRCUMDUCTION.
6. SADDLE JOINTS are bones which are saddle-shaped and move from side to side. This can be seen in the trapezium of the carpus and the metacarpal of the thumb.
7. BALL & SOCKET JOINTS contain one bone with a ball surface and another with a cup-like depression. These joints allow for ABDUCTION and ADDUCTION, ROTATION and CIRCUMDUCTION, FLEXION and EXTENSION. They can been illustrated by the hip and shoulder joints.

G. SPECIALIZED MOVEMENTS

1. The specialized movements that occur at diarthrotic joints in addition to flexion/extension, abduction/adduction, and rotation/circumduction are the following:
   - ELEVATION - upward movement of a body part

- DEPRESSION - downward movement of a body part
- PROTRACTION - movement of the mandible or shoulder forward
- RETRACTION - movement of the mandible or shoulder backward
- INVERSION - movement of the sole inward (medial)
- EVERSION - movement of the sole outward (lateral)
- DORSIFLEXION - movement of the foot upward
- PLANTARFLEXION - movement of the foot downward
- SUPINATION - forearm movement so that the palms are forward or up
- PRONATION - forearm movement so that the palms are backward or down

H. Knee Joint

1. The knee joint represents the most complex of the diarthrotic joints, and contains many structural features.
2. The main structures of the knee are as follows:
   - ARTICULAR CARTILAGE
   - PATELLAR LIGAMENT
   - OBLIQUE POPLITEAL LIGAMENT
   - ARCUATE POPLITEAL LIGAMENT
   - TIBIAL COLLATERAL LIGAMENT
   - FIBULAR COLLATERAL LIGAMENT
   - ANTERIOR CRUCIATE LIGAMENT
   - POSTERIOR CRUCIATE LIGAMENT
   - ANTERIOR MENISCI
   - MEDIAL MENISCUS
   - LATERAL MENISCUS
   - BURSAE

I. Common Disorders

1. RHEUMATISM refers to any painful state of supporting structures of the body such as bones, ligaments, tendons, joints, and muscles.
2. ARTHRITIS refers to several disorders characterized by inflammation of joints accompanied by stiffness.
3. RHEUMATOID ARTHRITIS (RA) is an autoimmune disease and is characterized by an inflammation of the synovial membranes.
4. OSTEOARTHRITIS is a degenerative joint disease characterized by deterioration of the articular cartilage. It is non-inflammatory and generally affects the weight-bearing joints.
5. GOUTY ARTHRITIS is a condition in which sodium ureate crystals are deposited in the soft tissues of joints, causing inflammation, swelling, and pain.
6. BURSITIS is an acute chronic inflammation of bursae, caused by trauma, infection, or rheumatoid arthritis.
7. A SPRAIN is a forcible wrenching or twisting of a joint with partial rupture to its attachments without dislocation.
8. A STRAIN is the stretching of a muscle beyond normal limits.

SUMMARY OF MAJOR JOINTS

COMMON JOINT	STRUCTURAL TYPE	FUNCTIONAL TYPE	MOVEMENT(S)
ANKLE	SYNOVIAL; HINGE	DIARTHROTIC	FLEXION; EXTENSION
ATLANTO-AXIAL	SYNOVIAL; PIVOT	DIARTHROTIC	ROTATION
ATLANTO-OCCIPITAL	SYNOVIAL; CONDYLOID	DIARTHROTIC	FLEXION; EXTENSION; ABDUCTION; ADDUCTION; CIRCUMDUCTION
COXAL	SYNOVIAL; BALL-AND-SOCKET	DIARTHROTIC	FLEXION; EXTENSION; ABDUCTION; ADDUCTION; ROTATION; CIRCUMDUCTION
ELBOW	SYNOVIAL; HINGE	DIARTHROTIC	FLEXION; EXTENSION
FINGERS	SYNOVIAL; HINGE	DIARTHROTIC	FLEXION; EXTENSION
GLENOHUMERAL	SYNOVIAL; BALL-AND-SOCKET	DIARTHROTIC	FLEXION; EXTENSION; AB/ADDUCTION; CIRCUMDUCTION; ROTATION
INTERVERTEBRAL	CARTILAGINOUS; SYMPHYSIS	AMPHIARTHROTIC	SLIGHT
KNEE	SYNOVIAL; HINGE	DIARTHROTIC	FLEXION; EXTENSION; SOME ROTATION
PUBIC SYMPHYSIS	CARTILAGINOUS; SYMPHYSIS	AMPHIARTHROTIC	SLIGHT
SACROILIAC	SYNOVIAL; PLANE	DIARTHROTIC	SLIGHT GLIDING TO NONE
SKULL	FIBROUS; SUTURE	SYNARTHROTIC	NONE
STERNOCLAVICULAR	SYNOVIAL	DIARTHROTIC	GLIDING
STERNOCOSTAL	CARTILAGINOUS; SYNCHONDROSIS	AMPHIARTHROTIC	SLIGHT
TIBIOFIBULAR-PROXIMAL	SYNOVIAL; PLANE	SYNARTHROTIC	GLIDING
TIBIOFIBULAR-DISTAL	FIBROUS; SYNDESMOSIS	SYNARTHROTIC	SLIGHT
TEMPERO-MANDIBULAR	SYNOVIAL	DIARTHROTIC	ELEVATION; DEPRESSION; PROTRACTION; RETRACTION
TOES	SYNOVIAL; HINGE	DIARTHROTIC	FLEXION; EXTENSION

IV. TEACHING TIPS AND SUGGESTIONS

A. HELPFUL HINTS

1. After the students have reviewed the various types of joints, have them display some movements. Students should then describe the bones involved, type of joint, and functional and structural classifications.
2. Using diarthrotic joints as a reference, demonstrate the difference between monoaxial, biaxial, and triaxial joints.
3. Models of the shoulder, hip, elbow, and knee joints are useful demonstration aids.

B. Essay Questions

1. For each of the following movements, indicate the bones that form the joints and the specific kind of joint involved:
   a. bending your elbow
   b. lowering your arm to the side
   c. lifting your leg laterally
   d. turning your head from side to side
   e. turning the plantar surface of the foot inward
   f. protruding your jaw

2. For each of the movements above, classify the joints as monoaxial, biaxial, triaxial, or non-axial.

C. Topics for Discussion

1. Discuss the stress placed on the ankles, knees and hips when running.
2. Discuss how lifting a heavy load affects the joints of the back, knees, and arms.

V. AUDIOVISUAL MATERIALS

A. Overhead Transparencies

1. Bone Joints (HSC).
2. PAP Transparency Set (Trs. 9.1a, 9.2a-f, 9.7a&b, c&e).

B. Videocassettes

1. Moving Parts (26 min.; FHS).

C. Films: 16 mm

1. Human Skeleton (16 min.; FHS).
2. Muscles and Joints (26 min.; FHS).
3. Our Wonderful Body: How It Moves (11 min.; 1968; COR/KSU).
4. Bionics: Man or Machine? (23 min.; 1977; EBEC/KSU).

D. Transparencies: 35 mm (2x2)

1. PAP Slide Set (Slides 30-31).
2. AHA Slide Set.
3. Pain in Rheumatology Medical Slides Series (188 Slides; CARO).

E. Computer Software

1. Bone Probe (Apple; SC-175027; PLP).
2. Skeletal System (Apple; SC-182005; IBM; SC-182006; PLP).
3. Body Language: Skeletal System Parts I-IV (EIL).

CHAPTER 8

MUSCULAR SYSTEM

CHAPTER AT A GLANCE

- Types of Muscle Tissue
- Functions of Muscle Tissue
- Characteristics of Muscle Tissue
- Skeletal Muscle Tissue
    - *Connective Tissue Components*
    - *Nerve and Blood Supply*
    - *Histology*
- Contraction
    - *Sliding Filament Mechanism*
    - *Neuromuscular Junction*
    - *Physiology of Contraction*
    - *Relaxation*
    - *Energy for Contraction*
    - *All-or-None Principle*
    - *Homeostasis*
        - *Oxygen Debt*
        - *Muscle Fatigue*
        - *Heat Production*
    - *Kinds of Contraction*
        - *Twitch*
        - *Tetanus*
        - *Isotonic and Isometric*
    - *Muscle Tone*
    - *Muscle Atrophy and Hypertrophy*
- Cardiac Muscle Tissue
- Smooth Muscle Tissue
- How Skeletal Muscles Produce Movement
    - *Origin and Insertion*
    - *Group Actions*
- Naming Skeletal Muscles
- Principle Skeletal Muscles
- Medical Terminology and Conditions
- Wellness Focus: Strength Training, Antidote for Aging

I. CHAPTER SYNOPSIS

This chapter deals with the physiological characteristics of muscle tissue and the classification of muscle tissue into skeletal, cardiac, and smooth muscle. The blood and nerve supplies of muscle are also considered. Attention is then directed to the ultrastructure of skeletal muscle and the physiology of muscle contraction. Emphasis is placed on the Sliding-Filament Theory, neuromuscular junctions, and the physiology of contraction. There is considerable discussion given to the energy for contraction and the maintenance of muscle homeostasis. Further discussion revolves around the functional role of the motor unit and the All-or-None Principle. The discussion continues with the types of skeletal muscle contractions including muscle twitch, treppe, tetanus, isotonic and isometric contractions. Muscle tone is then considered, with attention given to muscular atrophy and hypertrophy.

The relationship between bones and muscles is established. Movement by skeletal muscles is related to their origins, insertions, lever systems and group actions. Particular attention is given to the criteria used for naming muscles. The major skeletal muscles are then detailed by region and are accompanied by plate diagrams. The chapter concludes with a brief discussion of cardiac and smooth muscle, common muscular disorders, and a list of medical terms.

II. LEARNING GOALS/STUDENT OBJECTIVES

1. Describe the connective tissue components, blood and nerve supply, and histology of skeletal muscle tissue.
2. Explain the factors involved in the contraction of skeletal muscle.
3. Explain the relationship of muscle tissue to homeostasis.
4. Describe the various kinds of muscle contraction.
5. Define muscle tone and describe abnormalities related to it.
6. Describe the structure and function of cardiac muscle tissue.
7. Describe the structure and function of smooth muscle tissue.
8. Describe how skeletal muscles produce movement.
9. List and describe several ways that skeletal muscles are named.
10. For various regions of the body, describe the location of skeletal muscles and identify their functions.

III. SAMPLE LECTURE OUTLINE

A. TYPES OF MUSCLE TISSUE

1. The types of muscle are categorized by LOCATION, HISTOLOGY, and MODES OF CONTROL.
2. There are three recognizable types of muscle tissue: SKELETAL, CARDIAC and SMOOTH.
3. SKELETAL MUSCLE tissue is primarily attached to the bones. It is striated and under voluntary neural control.
4. CARDIAC MUSCLE is the principal muscle lining the walls of the heart. It is striated and under involuntary neural control.
5. SMOOTH MUSCLE, also known as visceral muscle, is located in the major organs (viscera) and the walls of the blood vessels. It is non-striated and under involuntary neural control.

B. FUNCTION OF MUSCLE TISSUE

1. Muscle tissue has four key functions. These include motion, movement of substances within the body, stabilizing body positions and regulating organ volume, and heat production.
2. MOTION in function is related to walking or running. It is also finely tuned as in grasping a pen.
3. MOVEMENT OF SUBSTANCES WITHIN THE BODY includes cardiac and smooth muscle. The cardiac muscle, through contraction, propels and moves the blood though the vessels of the body. Smooth muscle, through contraction, is involved in the movement of substances in the digestive, urinary and reproductive tracts.
4. STABILIZING BODY POSITION allows for proper posture during sitting and standing.
5. HEAT PRODUCTION arises from energy generated during contraction of muscle tissue.

C. CHARACTERISTICS OF MUSCLE TISSUE

1. Muscles are EXCITABLE- they are capable of receiving and responding to stimuli.
2. Muscles exhibit CONTRACTILITY- the ability to shorten and thicken.
3. Muscles exhibit EXTENSIBILITY- the ability to be stretched.
4. Muscles exhibit ELASTICITY- the ability to return to their original shape after contraction or extension.
5. The three MAJOR FUNCTIONS of muscle tissue are MOTION, MAINTENANCE OF POSTURE, and HEAT PRODUCTION.

D. SKELETAL MUSCLE TISSUE

1. Skeletal muscle tissue contains connective tissue components. The FASCIAE are sheets of fibrous connective tissue beneath the skin or around muscle and organs of the body. EPIMYSIUM, covers the entire muscle, PERIMYSIUM covers the FASCICULI, and ENDOMYSIUM covers the muscle fibers.
2. Extensions of connective tissue beyond the muscle cells include TENDONS and APONEUROSES. These attach muscle to bone or other muscles. TENDONS attach muscle to muscle and contain tendon sheaths which allow them to slide back and forth easily.
3. Skeletal muscle tissue contains both nerve and blood supplies. Nerves convey impulses for muscular contractions. Blood provides the essential nutrients and oxygen while removing many of the metabolic wastes resulting from contraction.
4. Histologically, skeletal muscle consists of FIBERS (cells) covered by a SARCOLEMMA (plasma membrane). The fibers contain SARCOPLASM, multiple nuclei, SARCOPLASMIC RETICULUM, and TRANSVERSE TUBULES.
5. Each muscle fiber contains smaller units called MYOFIBRILS that consist of thin and thick proteinaceous MYOFILAMENTS. The myofilaments are compartmentalized into units of contraction called SARCOMERES. The thin myofilaments are composed of ACTIN, TROPOMYOSIN and TROPONIN. The thick myofilaments are composed mostly of MYOSIN and contain projecting myosin heads called CROSS-BRIDGES that contain actin and ATP binding sites.

E. Histology

1. The mechanics of muscle contraction follow the SLIDING-FILAMENT THEORY.
2. A muscle ACTION POTENTIAL travels over the SARCOLEMMA and enters the TRANSVERSE TUBULES. This affects the SARCOPLASMIC RETICULUM and causes the release of calcium ions into the SARCOPLASM.
3. The MUSCLE ACTION POTENTIAL leads to the RELEASE OF CALCIUM ions from the SARCOPLASMIC RETICULUM which triggers the contractile response.
4. Actual contraction is brought about when the thin ACTIN MYOFILAMENTS of the sarcomere SLIDE toward one another and OVER THE THICK MYOSIN MYOFILAMENTS. During this process, the myosin cross-bridges pull on the actin myofilaments.
5. A MOTOR NEURON is responsible for transmitting the nerve impulse (action potential) to a skeletal muscle where it serves as the stimulus for contraction.
6. The point at which the motor neuron and the muscle sarcolemma meet is referred to as the NEUROMUSCULAR JUNCTION.
7. A chemical NEUROTRANSMITTER released at neuromuscular junctions in skeletal muscle is ACETYLCHOLINE (ACH).
8. The MOTOR (EFFERENT) NEURON and all of the muscle fibers it innervates form a MOTOR UNIT.
9. A single motor unit may affect as few as 10 or as many as 2000 muscle fibers although the average is approximately 150.
10. The number of motor units that fire in a muscle at any one time is the basis for the variability in contraction.
11. The process of increasing the number of motor units firing at any one time is called RECRUITMENT.

F. Skeletal Muscle Contraction

1. When a nerve impulse, or ACTION POTENTIAL, reaches the AXON TERMINAL, the SYNAPTIC VESICLES in the axon terminal release AcH.
2. The release of AcH ultimately initiates a muscle ACTION POTENTIAL in the muscle fiber SARCOLEMMA.
3. The muscle ACTION POTENTIAL travels into the TRANSVERSE TUBULES and causes the SARCOPLASMIC RETICULUM to release STORED CALCIUM ions into the SARCOPLASM.
4. The released calcium combines with TROPONIN which pulls on the tropomyosin filaments and changes their orientation. This exposes the myosin-binding sites on the actin myofilament.
5. Splitting ATP with ATPASE into ADP+ phosphate releases energy which activates the myosin cross-bridges.
6. The ACTIVATED CROSS-BRIDGES attach to the actin myofilament, and a change in the orientation of the cross-bridges occurs. This is called a POWER STROKE.
7. The POWER STROKE results in the sliding of the thin actin myofilaments.
8. Repeated detachment and reattachment of the cross-bridges results in the shortening (ISOTONIC CONTRACTION) or increased tension without shortening (ISOMETRIC CONTRACTION) of the muscle.
9. The enzyme, acetylcholinesterase (AcHE), in the neuromuscular junction, destroys acetylcholine and stops the generation of a muscle action potential.
10. Calcium ions are then reabsorbed into the sarcoplasmic reticulum, exposing the troponin. This causes the cross-bridges to separate and the muscle fiber resumes its resting state.

G. ENERGY NEEDED FOR CONTRACTION

1. The immediate and direct source of energy for muscle contraction is ATP.
2. PHOSPHOCREATINE and the metabolism of GLYCOGEN and fats are necessary for the continuous generation of ATP by muscle fibers.
3. PHOSPHOCREATINE and ATP comprise the PHOSPHAGEN SYSTEM, which allows for initial contraction and short bursts of maximal contraction lasting up to fifteen seconds.
4. The GLYCOGEN-LACTIC ACID SYSTEM provides energy via glycolysis. This is an anaerobic process which will allow thirty to forty seconds of maximum muscular contraction after the phosphocreatine supply is exhausted.
5. An AEROBIC RESPIRATION and GLYCOLYSIS are used for prolonged muscular contraction and will function efficiently as long as oxygen and nutrients are present in adequate amounts.

H. THE ALL-OR-NONE PRINCIPLE

1. The weakest stimulus capable of causing contraction in a motor unit is called a LIMINAL or THRESHOLD STIMULUS.
2. A stimulus not capable of inducing contraction is called a SUBLIMINAL or SUBTHRESHOLD STIMULUS.
3. The ALL -OR -NONE PRINCIPLE states that once a threshold or liminal stimulus is applied, all of the individual muscle fibers of a motor unit will contract to their greatest extent. Any further increase in the degree of stimulation will not cause a corresponding increase in muscle contraction.

I. TYPES OF SKELETAL MUSCLE CONTRACTION

1. Skeletal muscles are capable of producing several different types of contractions, depending upon the strength and frequency of stimulation.
2. The various kinds of contractions are TWITCH, TETANUS, TREPPE, ISOTONIC, and ISOMETRIC.
3. A MYOGRAM is a record of muscular contraction shows the various phases involved in the various phases of muscular contraction.
4. The three major phases are the LATENT PERIOD, CONTRACTION PERIOD, and RELAXATION or REFRACTORY PERIOD.
5. The LATENT PERIOD is the time lapse between the action potential and the response of the muscle fibers.
6. The CONTRACTION PERIOD is the time frame in which the muscle fibers are undergoing shortening.
7. The REFRACTORY PERIOD is that time frame when a muscle has temporarily lost its excitability.
8. Skeletal muscles have short refractory periods.
9. WAVE SUMMATION results in an increase in the strength of muscle contraction. It occurs when a second stimulus is applied to the muscle before it has completely relaxed after the previous stimulus.
10. A sustained partial contraction of portions of skeletal muscle results in muscle tone and occurs even in relaxed or uncontracted muscle.
11. MUSCLE TONE is essential for the maintenance of posture.
12. HYPOTONIA is a decrease in muscle tone. HYPERTONIA is an increase in muscle tone.
13. MUSCULAR ATROPHY refers to the wasting away of muscle. Muscular hypertrophy refers to the increase in the diameter of muscle fibers, resulting from enlargement or overgrowth.

J. Skeletal Muscles and Movement

1. Skeletal muscles produce movements by exerting force on tendons, which in turn, pull bones or other structures, such as skin.
2. Most muscles cross at least one joint and are attached to the articulating bones that form the joint. The bones serve as levers while the joints serve as fulcrums.
3. When a muscle contracts, it draws a movable bone toward a stationary one.
4. The ORIGIN is the point of attachment on the stationary bones, while the INSERTION is the point of attachment on movable bones.
5. Skeletal muscle fibers (cells) are arranged in a parallel fashion, in bundles called FASCICULI.
6. The FASCICULI are arranged relative to the tendons in one of four arrangements: parallel, convergent, pennate, or circular.
7. Fascicular arrangement can be correlated to muscle strength and range of movement.

K. Cardiac Muscle

1. CARDIAC MUSCLE is found only in the HEART. It is striated and neurologically involuntary.
2. The muscle fibers are quadrangular and generally contain a single, centrally located nucleus.
3. Compared to skeletal muscle, cardiac muscle contains more sarcoplasm, mitochondria, larger transverse tubules, and a less developed sarcoplasmic reticulum.
4. The myofilaments are not arranged in discrete myofibrils.
5. Muscle fibers branch freely and are connected via gap junctions.
6. Cardiac muscle cells are connected via INTERCALATED DISCS which aid in the conduction of muscle action potentials by way of the gap junctions located at the discs.
7. Cardiac muscle contracts and relaxes rapidly, continuously, and rhythmically.
8. Energy for contraction is supplied by large, numerous mitochondria and through the catabolism of fats and glycogen.
9. Cardiac muscle is MYOGENIC, it can contract without extrinsic stimulation and can remain contracted for sustained periods of time.
10. Cardiac muscle exhibits a long refractory period, referred to as a PLATEAU POTENTIAL, which prevents tetanus of the muscle.

L. Smooth Muscle

1. SMOOTH MUSCLE is non-striated and involuntary. It is also referred to as VISCERAL muscle.
2. The smooth muscle fibers are elongated, tapered fibers with a single, centrally located nucleus. They contain actin and myosin myofilaments, but unlike skeletal muscle, the filaments are not arranged in an orderly fashion.
3. VISCERAL or single unit smooth muscle, is found in the walls of the viscera as well as the walls of both small arteries and veins.
4. MULTIUNIT SMOOTH MUSCLE is found in large blood vessels, large airways, and in the eyes. The muscle fibers operate singly, rather than as a unit.
5. Compared to skeletal muscle, smooth muscle displays a longer duration of contraction and relaxation.
6. Smooth muscle fibers respond to nerve impulses, hormones, and local chemical factors.
7. Smooth muscle fibers are very extensible without developing tension.

M. Group Actions of Skeletal Muscle

1. Most movements are coordinated by several skeletal muscles acting in groups rather than individually.
2. Most skeletal muscles are arranged in opposing pairs at joints.
3. The muscle that causes a desired movement is called the AGONIST; the ANTAGONIST produces the opposite action.
4. Most movements also involve muscles called SYNERGISTS. These function to steady movements and to aid the agonist in functioning more efficiently.
5. Muscles can also be FIXATORS which stabilize the origin of the agonist so that it can operate more efficiently.
6. Most muscles can act as AGONISTS, ANTAGONISTS, SYNERGISTS, or FIXATORS depending on the movement.

N. Naming Skeletal Muscles

1. There are nearly 700 skeletal muscles. Their names are based on several characteristics. The muscle names may indicate the DIRECTION of MUSCLE FIBERS or their SIZE, SHAPE, ACTION, SITE of ORIGIN, or ORIGIN, and INSERTION POINTS.
2. Reference should be made to Exhibit 8.2 on Naming Skeletal Muscles for a detailed listing.

PRINCIPAL SKELETAL MUSCLES

FACIAL EXPRESSION	LOWER JAW	EYEBALL MOVEMENT
EPICRANIUS	MASSETER	SUPERIOR RECTUS
FRONTALIS	TEMPORALIS	INFERIOR RECTUS
OCCIPITALIS		LATERAL RECTUS
ZYGOMATICUS MAJOR		MEDIAL RECTUS
BUCCINATOR		SUPERIOR OBLIQUE
PLATYSMA		INFERIOR OBLIQUE
ORBICULARIS OCULI		

ANTERIOR ABDOMEN	BREATHING	SHOULDER GIRDLE
RECTUS ABDOMINIS	DIAPHRAGM	SUBCLAVIUS
EXTERNAL OBLIQUE TRANSVERSUS ABDOMINIS	EXTERNAL & INTERNAL PECTORALIS MINOR	RHOMBOIDEUS MAJOR PECTORALIS MINOR
INTERNAL OBLIQUE	INTERCOSTALS	SERRATUS ANTERIOR
		TRAPEZIUS
		LEVATOR SCAPULAE

ARM MOVEMENT	FOREARM MOVEMENT	VERTEBRAL COLUMN
PECTORALIS MAJOR	BICEPS BRACHII	STERNOCLEIDOMASTOID
LATISSIMUS DORSI	BRACHIALIS	RECTUS ABDOMINIS
DELTOID	TRICEPS BRACHII	QUADRATUS LUMBORUM
SUBSCAPULARIS	SUPINATOR	SACROSPINALIS
SUPRASPINATUS	PRONATOR TERES	
INFRASPINATUS		
TERES MAJOR		

THIGH MOVEMENT	LOWER LEG MOVEMENT
PSOAS MAJOR	ADDUCTOR MAGNUS
ILIACUS	ADDUCTOR LONGUS
GLUTEUS MAXIMUS	PECTINEUS
GLUTEUS MEDIUS	GRACILIS
GLUTEUS MINIMUS	QUADRICEPS FEMORIS
TENSOR FACIA LATA	SARTORIUS
ADDUCTOR LONGUS	BICEPS FEMORIS
ADDUCTOR MAGNUS	SEMITENDINOSUS
PECTINEUS	SEMIMEMBRANOSUS

IV. TEACHING TIPS AND SUGGESTIONS

A. Helpful Hints

1. The section on microstructure is often very confusing and bears going over very slowly with numerous visual aids. Try to have the students visualize the whole muscle and then break it down into the successively smaller sarcomeric components.
2. Incorporate a visual aid to illustrate the events associated with the contraction and relaxation of smooth muscle.
3. In discussing the motor unit, be sure to point out that full contraction of the fibers of a single motor unit does not mean the full contraction of the entire muscle. Each muscle has many motor units which do not contract at the same time. They can be added together to present a graded contraction to perform different tasks.
4. It is often useful to compare the structural and functional differences and similarities between the three types of muscle.
5. If appropriate, supplement dissection with the use of a live model.

B. Essay Questions

1. What is the relationship between low levels of AcH and an individual who exhibits droopy eyelids, poor muscle tone, flaccid skeletal muscles, and speech problems?
2. A long-distance runner is about to enter a ten kilometer race. She spends several minutes warming up. After the race she is out of breath and collapses after crossing the finish line. What type of muscle contraction is involved in warming up? Why was the runner out of breath after the race? What was happening while she was attempting to catch her breath?
3. Flex your leg at the knee joint. Determine which muscles are the agonists and which are the antagonists. What muscles are the synergists for the flexing action?

V. AUDIOVISUAL MATERIALS

A. Videocassettes

1. The Physiology of Exercise (15 min.; 1988; FHS).
2. Muscle Power (26 min.; FHS).
3. What Can Go Wrong? (40 min.; PLP).
4. Oh, My Aching Back! (28 min.; 1985; KSU).
5. Sports Injuries (28 min.; 1985; KSU).

B. Films: 16 mm

1. Muscles: A Study of Integration (25 min.; 1972; KSU).
2. Muscles (30 min.; 1961; McG/KSU).
3. The Human Body: Muscular System (14 min.; 1980; COR/KSU).
4. Muscle Power (26 min.; FHS).
5. Muscle (30 min.; CRM).
6. Muscle: Electrical Activity of Contraction (9 min.; 1969; EBEC).

7. Muscle: Dynamics of Contraction (21 min.; 1969; EBEC/KSU).
8. Muscle: Chemistry of Contraction (15 min.; 1969; EBEC/KSU).
9. The Muscular Spindle (19 min.; 1970; UIFC/MG/KSU).
10. Myoglobin (10 min.; 1971; SB).
11. What Makes Muscle Pull? (8 min.; 1971; MG/KSU).

C. TRANSPARENCIES: 35 MM (2x2)

1. PAP Slide Set (Slides 32-28).
2. AHA Slide Set.
3. Muscular System (Slides 32-46; McG).
4. Muscular System and Its Function (20 Slides; EIL).
5. What Is a Muscle Set? (53 Slides; CARO).
6. Muscular Tissue Set (20 Slides; CARO).

D. OVERHEAD TRANSPARENCIES

1. Muscle Contraction (20 Transparencies; EIL).
2. The Skeleton, Muscles, and Internal Organs (1 Transparency with 3 overlays; CARO).
3. Human Muscles I-II (1 Transparency each; CARO).
4. The Structure and Antagonism of Muscles (1 Transparency; CARO).
5. Skeleton: Muscles: Back View; Skeleton: Muscles: Front View; (GAF).
6. Muscular System- Unit 4 (13 Transparencies RJB).
7. PAP Transparency Set (Trs. 10.1, 10.3b, 10.3c&d, 10.4a-e, 10.6a-c, 10.7, 10.8a&b, 10.9, 10.10, 10.11a-c, 10.12, 10.13, 10.14, 10.17a&b, 10.18a-c).

E. COMPUTER SOFTWARE

1. Muscular System (Apple; SC-182007; IBM; SC-182008; PLP).
2. Skeletal Muscle Anatomy and Physiology (Apple; SC-510001; PLP).
3. Neuromuscular concepts (Apple; SC-510005; PLP).
4. Body Language: Muscular System I-III (Apple II Series; FHS).

CHAPTER 9

NERVOUS TISSUE

CHAPTER AT A GLANCE

- ORGANIZATION
- HISTOLOGY
  - *Neuroglia*
  - *Neurons*
    - *Structure*
    - *Formation of Myelin*
    - *Grouping of Neural Tissue*
- FUNCTIONS
  - *Nerve Impulse*
    - *Ion Channels in Plasma Membrane*
    - *Membrane Potentials*
    - *Excitability*
    - *All-or-None Principle*
    - *Continuous and Saltatory Conduction*
    - *Speed of Nerve Impulses*
  - *Conduction Across Synapses*
  - *Regeneration of Nervous Tissue*
- WELLNESS FOCUS: THE CONSTRICTION OF ADDICTION

I. CHAPTER SYNOPSIS

After identifying the organization of the nervous system into its subdivisions of the CENTRAL NERVOUS SYSTEM (CNS), AUTONOMIC NERVOUS SYSTEM (ANS) and PERIPHERAL NERVOUS SYSTEM (PNS), the histology of the supportive neuroglial and functional neuron cells is considered. Attention is given to the components of a neuron, and their structural and functional classification. The physiology of nervous tissue is then discussed. The topic areas considered are plasma membrane ion channels, membrane potentials, excitability, the All-or-None Principle, saltatory conduction and speed of nerve impulses. The physiology of impulse conduction synapses are then considered. Particular attention is given to excitatory and inhibitory transmission, synaptic integration and alternation of synaptic conduction. The chapter concludes with a discussion of graded potentials and neural regeneration.

II. LEARNING GOALS/STUDENT OBJECTIVES

1. Describe the organization of the nervous system.
2. Compare the structure and fucntions of neuroglia and neurons.
3. Describe how a nerve impulse is generated and conducted.

III. SAMPLE LECTURE OUTLINE

A. ORGANIZATION OF NERVOUS TISSUE

1. The nervous and endocrine systems control and integrate all body activities and aid in maintaining homeostasis.
2. The basic functions of the nervous system are to sense changes (SENSORY), to interpret these changes (INTEGRATIVE), and to react to changes (MOTOR).
3. The CENTRAL NERVOUS SYSTEM consists of the brain and spinal cord.
4. The PERIPHERAL NERVOUS SYSTEM is classified into the AFFERENT (sensory) system and the EFFERENT (motor) system.
5. The AFFERENT SYSTEM includes all receptors as well as some sensory neurons.
6. The EFFERENT SYSTEM is divided into the SNS and the ANS.
7. The SNS consists of efferent neurons that conduct impulses from the central nervous system to skeletal muscles.
8. The ANS contains efferent neurons that transmit impulses from the central nervous system to smooth muscle tissue, cardiac muscle tissue, and glands.

B. HISTOLOGY

1. NEUROGLIA are specialized tissue cells that support NEURONS, attach neurons to blood vessels, produce the myelin sheath around axons of the CNS, and carry on phagocytosis.
2. The NEUROGLIAL CELLS include ASTROCYTES, OLIGODENDROCYTES, MICROGLIA and EPENDYMA.
3. The FUNCTIONAL CELL of the nervous system, the NEURON, consists of three distinct portions. These are the CELL BODY, DENDRITES, and the AXON.
4. The CELL BODY (soma or perikaryon) contains a well-defined single nucleus, a nucleolus, and granular cytoplasm.
5. The DENDRITES convey impulses to the cell body. Neurons may have many dendrites.
6. The AXON conducts nerve impulses from the neuron to the dendrites, the cell body of another neuron or an effector organ of the body such as a muscle or a gland. Each neuron contains only one axon.
7. Axons, especially those outside of the CNS, are often covered with a white phospholipid material called MYELIN. Non-myelinated neurons lack this material, which functions to increase the speed of impulse conduction as well as to insulate and maintain the axon.
8. Neurons are classified on the basis of structure which is determined by the number of processes extending from the cell body. They can be classified as MULTIPOLAR, BIPOLAR, or UNIPOLAR neurons.
9. Neurons are also classified on the basis of function and by the direction in which the information is transmitted. SENSORY NEURONS (afferent) convey information from receptors to the CNS. ASSOCIATION NEURONS (interneurons) conduct impulses to other neurons, and MOTOR (EFFERENT) NEURONS conduct impulses to effectors such as muscles or glands.
10. The processes of afferent and efferent neurons are arranged into bundles called NERVES outside of the CNS or FIBER TRACTS within the CNS.

C. MEMBRANE POTENTIAL AND EXCITABILITY

1. The NERVE IMPULSE or nerve ACTION POTENTIAL is the body's quickest way of controlling and maintaining homeostasis.

2. The membrane of a resting (non-conducting) neuron is positively charged outside and negatively charged inside. This is partly due to the permeability properties of the membrane toward sodium and potassium ions and negatively charged organic phosphate and proteins.
3. When a LIMINAL or THRESHOLD STIMULUS is presented to the neuron, the inside of the neuron cell membrane becomes positively charged and the outside becomes negatively charged.
4. The membrane is said to be DEPOLARIZED, and an ACTION POTENTIAL is produced, which travels from point to point along the membrane. The traveling action potential is the NERVE IMPULSE.
5. The ABILITY OF A NEURON TO RESPOND to a stimulus and convert it into a nerve impulse is called EXCITABILITY.
6. RESTORATION of the resting membrane potential is called REPOLARIZATION.
7. The period of time during which the membrane recovers and cannot initiate another action potential is called the ABSOLUTE REFRACTORY PERIOD.
8. According to the ALL-OR-NONE PRINCIPLE, if a stimulus is of a liminal or threshold value, the impulse will travel the entire length of the neuron at a constant and maximum strength.
9. Nerve impulse conductions which jump from neurofibril node to neurofibril node is called SALTATORY CONDUCTION, and is found principally in myelinated fibers.
10. Impulse conduction in fibers without myelin is called NONDECREMETAL CONDUCTION or CONTINUOUS CONDUCTION, and is significantly slower than saltatory conduction.
11. The speed of impulse conduction is independent of the stimulus strength. Nerve fibers with large diameters conduct impulses faster than those with smaller diameters; myelinated fibers conduct impulses faster than unmyelinated fibers; and warm nerve fibers conduct impulses faster than cooler nerve fibers.

D. CONDUCTION ACROSS SYNAPSES

1. Nerve impulse conduction occurs not only along the length of a neuron, but also from one neuron to another, or from a neuron to an effector.
2. The junction between neurons is called a SYNAPSE.
3. At a chemical synapse, there is only ONE-WAY nerve impulse conduction to a postsynaptic dendrite, cell body, or axon hillock.
4. SYNAPTIC TRANSMISSION is either EXCITATORY or INHIBITORY.
5. An EXCITATORY TRANSMITTER-RECEPTOR INTERACTION is one that can depolarize or lower the postsynaptic neuron's membrane potential, so that new impulses can be generated across the synapse.
6. An INHIBITORY TRANSMITTER-RECEPTOR INTERACTION is one that can increase the postsynaptic neuron's membrane potential, so that new impulses are impeded from being generated across the synapse.
7. FACILITATION refers to a state of near-threshold excitation, so that subsequent stimuli can generate an impulse more efficiently.
8. If several presynaptic end-bulbs release their neurotransmitter at about the same time, the combined effect may generate a nerve impulse, resulting in a phenomenon called SUMMATION.
9. The time for neurotransmitters to cross the synaptic cleft and have their effect on the postsynaptic neuron is approximately 0.5 milliseconds. This time frame is called SYNAPTIC DELAY.
10. An INHIBITORY TRANSMITTER-RECEPTOR INTERACTION is one that can raise the postsynaptic neuron's membrane potential and thus inhibit or prevent an impulse at the synapse.
11. PRESYNAPTIC INHIBITION is another mechanism by which a nerve impulse in inhibited from crossing the synapse.

12. In all instances, the postsynaptic neuron acts as an integrator. It receives signals; integrates them and responds accordingly.

E. Neurotransmitters

1. A neurotransmitter which causes excitation in a major portion of the central nervous system is ACh.
2. Neurotransmitters that are inhibitory are gamma aminobutyric acid (GABA) and glycine.

F. Neuron Regeneration

1. Approximately at the age of six months, the neuron cell bodies of most developing nerve cells lose their mitotic apparatus and are no longer able to divide. Such neurons are incapable of replenishing themselves.
2. Nerve fibers that possess a neurilemma are capable of regeneration in most instances.

IV. TEACHING TIPS AND SUGGESTIONS

A. Helpful Hints

1. Indicate that the white color of most nerves, the spinal cord, and the inside of the brain is due to myelinated fibers.
2. Since myelination is incomplete at birth and continues through childhood, certain functions of the nervous system are modified during that time period. One such example is coordination of movements. Another is voluntary control of external urethral sphincter muscle at the outlet of the urinary bladder. Myelination of the nerves which control this function are usually complete by the third year of life.
3. In detailing impulse conduction it is often helpful to assign the corresponding sections of the chapter at least two periods prior to your discussion. Ask the students to list a sequential synopsis of the events surrounding the initiation of conduction of a nerve impulse.
4. It is useful to remind students that muscle fibers are also capable of producing action potentials and that they are not analogous to nerve fibers.
5. Indicate the differences in the All-or-None Principle as it applies to muscles and nerves.
6. Differentiate carefully the differences between electrical and chemical synapses.

B. Essay Questions

1. Prepare a labelled diagram illustrating the initiation of an impulse in a presynaptic neuron and the conduction across the synapse to a postsynaptic neuron.
2. Explain the differences in impulse conduction between A, B, and C nerve fibers, detailing the functions of each type with a specific example and noting any possible advantages. How does this relate to homeostasis?

C. Topics for Discussion

1. Discuss how the application of a cold compress to an injury can substantially reduce pain sensations. Describe what allows for the blockage of pain sensations from the injury site to the central nervous system.

V. AUDIOVISUAL MATERIALS

A. Videocassettes

1. The Nature of the Nerve Impulse (15 min.; 1988; FHS).
2. Nerves At Work (26 min.; HRM).
3. The Nervous System (48 min.; FHS).
4. The Nature of the Nerve Impulse (15 min.; FHS).
5. Staying Off Cocaine (38 min.; HRM).
6. Cocaine and Crack: Formula for Failure (21 min.; HRM).

B. Films: 16 mm

1. Fundamentals of the Nervous System (17 min.; EBEC).
2. The Nerve Impulse (22 min.; 1970; EBEC/KSU/UIFC).
3. The Human Body: Nervous System (22 min.; 1980; KSU/COR).
4. Signals and Receptors (24 min.; 1977; KSU).

C. Transparencies: 35 mm (2x2)

1. PAP Slide Set (Slides 43-48).
2. AHA Slide Set.
3. Visual Approaches to Histology: Nervous System (FAD).
4. Histology of the Nervous System (EIL).
5. Nervous System and Its Functions (20 Slides; EIL).
6. Nervous Tissue Set (20 Slides, CARO).

D. Overhead Transparencies

1. PAP Transparency Set (Trs. 12.1a, 12.3a, 12.4a&b, 12.5a-c, 12.8, 12.9a&b, 12.10, 12.11, 12.12, 12.13a, 12.14, 12.15, 12.16 & 12.18).
2. IM Transparency Masters (TM4).
3. The Central Nervous System (1 Transparency with 2 overlays; CARO).

E. Computer Software

1. Nervous System (Apple; SC-182013; IBM; SC-182014; PLP).
2. The Human Brain: Neurons (Apple; SC-510003; PLP).
3. A & P: The Human Brain: Neurons (Apple; PLP).
4. Body Language and the Nervous System (Apple; IBM; MAC; PLP).
5. Flash: Neurons (Apple; IBM; PLP).

6. Nervous System (Apple; IBM; PLP).
7. The Body Electric (Apple IIs; IBM; PC; MAC; Queue).
8. The Human Brain: Neurons (Apple; PLP).
9. Understanding Human Physiology: Membrane Potentials (IBM; PLP).

CHAPTER 10

CENTRAL AND SOMATIC NERVOUS SYSTEMS

CHAPTER AT A GLANCE

- Spinal Cord
  - *Protection and Coverings*
    - *Vertebral Canal*
    - *Meninges*
  - *General Features*
  - *Structure in Cross Section*
  - *Functions*
    - *Impulse Conduction*
    - *Reflex Center*
    - *Reflex Arc and Homeostasis*
- Spinal Nerves
  - *Names*
  - *Composition and Coverings*
  - *Distribution*
    - *Branches*
    - *Plexuses*
    - *Intercostal (Thoracic) Nerves*
- Brain
  - *Principal Parts*
  - *Protection and Coverings*
  - *Cerebrospinal Fluid (CSF)*
  - *Blood Supply*
  - *Brain Stem*
    - *Medulla Oblongata*
    - *Pons*
    - *Midbrain*
  - *Diencephalon*
    - *Thalamus*
    - *Hypothalamus*
  - *Reticular Activating System- Consciousness and Sleep*
  - *Cerebrum*
    - *Lobes*
    - *Brain Lateralization (Split Brain Concept)*
    - *White Matter*
    - *Basal Ganglia (Cerebral Nuclei)*
    - *Limbic System*
    - *Functional Areas of Cerebral Cortex*
    - *Memory*

- *Electroencephalogram (EEG)*
- *Cerebellum*
- NEUROTRANSMITTERS
- CRANIAL NERVES
- COMMON DISORDERS
- MEDICAL TERMINOLOGY AND CONDITIONS
- WELLNESS FOCUS: IN SEARCH OF MORPHEUS

I. CHAPTER SYNOPSIS

This chapter initially considers the structural groupings of neural tissue and the principal anatomical and functional features of the spinal cord, its coverings, the meninges. Next the spinal cord is discussed in terms of its functions as a conduction pathway and a reflex center. The important reflexes are categorized according to type, and several clinically important reflexes are highlighted. In addition, the spinal nerves and plexuses are covered. This is followed by the principal anatomical and functional features of the brain, cranial meninges, blood supply and the formation of cerebrospinal fluid. Also discussed are brain lateralization, neurotransmitters in the brain, and the location and functions of the cranial nerves. Disorders of the nervous system such as cerebrovascular accident (CVA), epilepsy, transient ischemic attack (TIA), poliomyelitis, multiple sclerosis, dyslexia Tay-Sacs disease, headache, and Reye's syndrome are covered. The chapter concludes with a list of medical terms and conditions.

II. LEARNING GOALS/STUDENT OBJECTIVES

1. Describe how the spinal cord is protected.
2. Describe the structure and functions of the spinal cord.
3. Describe the composition, coverings, and branches of a spinal nerve.
4. Discuss how the brain is protected and supplied with blood, and name the principal parts of the brain and explain the function of each part.
5. Explain the functions of selected neurotransmitters.
6. Identify the 12 pairs of cranial nerves by name, number, type, location, and function.

III. SAMPLE LECTURE OUTLINE

A. SPINAL CORD: GENERAL FEATURES AND CROSS-SECTION STRUCTURE

1. The SPINAL CORD is protected by the bony vertebral column, meninges, cerebrospinal fluid, and vertebral ligaments.
2. The MENINGES are the coverings that run continuously around the brain and spinal cord. They include the OUTER DURA MATER, THE MIDDLE ARACHNOID, and the INNER PIA MATER.
3. The spinal cord begins as a continuation of the medulla oblongata and TERMINATES at about the second lumbar vertebra. It contains cervical and lumbar enlargements that serve as points of origin for nerves to the extremities.
4. The tapered portion of the spinal cord is the CONUS MEDULLARIS, from which the FILUM TERMINALE and CAUDA EQUINA arise.

5. The spinal cord is partially divided into right and left sides by the ANTERIOR MEDIAN FISSURE and POSTERIOR MEDIAN SULCUS.
6. THE GRAY COMMISSURES, CENTRAL CANAL, ANTERIOR, POSTERIOR, and LATERAL GRAY HORNS, LATERAL WHITE COLUMNS, and ASCENDING and DESCENDING TRACTS can be observed in cross-section.
7. The CENTRAL CANAL, runs the length of the spinal cord and contains CEREBROSPINAL FLUID.
8. The GRAY MATTER of the spinal cord is divided into HORNS, while the WHITE MATTER is divided into COLUMNS.
9. The spinal cord conducts sensory and motor information via ascending and descending tracts, respectively.

B. GROUPING OF NEURAL TISSUE

1. WHITE MATTER refers to aggregations of myelinated processes from many neurons supported by associated neuroglia or Schwann cells.
2. GRAY MATTER is a collection of unmyelinated nerve cell bodies and dendrites or unmyelinated axons and terminal neuroglia.
3. A NERVE is a bundle of nerve axons and/or dendrites located outside of the central nervous system.
4. A GANGLION is a collection of unmyelinated neuron cell bodies that lie outside of the central nervous system.
5. A TRACT is a bundle of myelinated fibers of similar function in the central nervous system. They may run long distances up or down the spinal cord. They also exist in the brain and connect parts of the brain with one another or with the spinal cord.
6. ASCENDING TRACTS carry sensory information up the spinal cord toward the brain.
7. DESCENDING TRACTS conduct motor impulses from the brain down the spinal cord and to the spinal nerves.
8. A NUCLEUS is a mass of unmyelinated nerve cell bodies and dendrites, which is present as gray matter in the brain and spinal cord.
9. A HORN OR COLUMN is an area of gray matter in the spinal cord.

C. IMPULSE CONDUCTION ALONG TRACTS

1. Sensory information is transmitted from receptors up the spinal cord to the brain. Nerve conduction is along two general pathways on either side of the cord. These are the ANTERO-LATERAL (SPINOTHALAMIC) PATHWAY and the POSTERIOR COLUMN - MEDIAL LEMNISCUS PATHWAY.
2. The anterolateral pathway is comprised of the ANTERIOR SPINOTHALAMIC and LATERAL SPINOTHALAMIC, which convey impulses for sensory information such as pain, temperature, crude touch, pressure, tickle, and itch.
3. The posterior column medial lemniscus pathway, which is comprised of the FASCICULUS GRACILIS and FASCICULUS CUNEATUIS, convey nerve impulses for sensory proprioception, discriminating touch, stereognosis, weight discrimination, and vibration.
4. After sensory perception and interpretation is accomplished by higher centers, sensory-motor integration occurs.
5. After integration from selected areas of the brain output is sent down the spinal cord in two major descending pathways: the DIRECT (PYRAMIDAL) PATHWAYS and INDIRECT (EXTRAPYRAMIDAL) PATHWAYS.

6. The direct pathways include the LATERAL CORTICOSPINAL, ANTERIOR CORTICOSPINAL, and CORTICOBULBAR TRACTS. These convey impulses which cause precise voluntary movements of skeletal muscles.
7. The indirect pathways include the RUBROSPINAL, TECTOSPINAL, and VESTIUBULOSPINAL TRACTS. These relay impulses that program autonomic movements, help coordinate body movements with visual stimuli, maintain skeletal muscle tone and pressure, and regulate muscle tone in response to movements of the head.

D. SPINAL CORD FUNCTIONS

1. The major function of the spinal cord is to CONVEY SENSORY INFORMATION from the periphery to the brain and to CONDUCT MOTOR IMPULSES from the brain to the periphery via ascending and descending tracts.
2. The spinal cord serves as a REFLEX CENTER, which structurally involves the POSTERIOR ROOT, POSTERIOR ROOT GANGLION and the ANTERIOR ROOT OF THE SPINAL CORD.
3. A REFLEX ARC is the shortest route that can be taken by an impulse from a receptor to an effector. Its basic components are a RECEPTOR, a SENSORY NEURON, a CENTER which contains one or more ASSOCIATION NEURONS, a MOTOR NEURON, and an EFFECTOR.
4. A REFLEX is a quick, involuntary response to a stimulus that passes through the reflex arc. Reflexes represent the body's principal mechanisms for responding to certain changes in the internal and external environments.
5. The SOMATIC SPINAL REFLEXES include the stretch reflex, tendon reflex, flexor reflex, and crossed extensor reflex; all of which exhibit reciprocal innervation.
6. Reflexes can be classified as MONOSYNAPTIC REFLEXES or POLYSYNAPTIC REFLEXES containing two or more neurons, respectively.
7. Clinically important reflexes include the patellar reflex, the Achilles reflex, the Babinski reflex, and the abdominal reflex.

E. SPINAL NERVES

1. The 31 pairs of SPINAL NERVES are named and numbered according to the region and level of the spinal cord from which they emerge.
2. There are 8 pairs of CERVICAL, 12 pairs of THORACIC, 5 Pairs of LUMBAR, 5 pairs of SACRAL, and 1 pair of COCCYGEAL NERVES.
3. Spinal nerves are attached to the spinal cord by means of a POSTERIOR ROOT and an ANTERIOR ROOT. All spinal nerves are MIXED, they carry both sensory and motor information.
4. Sensory information enters the spinal cord via the DORSAL ROOT and motor impulses leave the spinal cord via the ANTERIOR ROOT.
5. Spinal nerves are covered by an ENDONEURIUM, PERINEURIUM, and an EPINEURIUM.
6. After a spinal nerve exits the INTERVERTEBRAL FORAMEN it divides into several branches called RAMI.
7. The BRANCHES of a spinal nerve include the DORSAL RAMUS, VENTRAL RAMUS, MENINGEAL BRANCH and the RAMI COMMUNICANTES.
8. The VENTRAL RAMI of SPINAL NERVES except for T2-T11 form networks of nerves called PLEXUSES. Emerging from the plexuses are nerves bearing names that are often descriptive of the general regions that they supply and the course they will take.
9. The CERVICAL PLEXUS supplies the skin and muscles of the head, neck and upper part of the shoulders, connects with some of the cranial nerves, and supplies the diaphragm.

10. The BRACHIAL PLEXUS makes up the nerve supply for the upper extremities and a number of neck and shoulder muscles.
11. The LUMBAR PLEXUS supplies the anterolateral abdominal wall, external genitals, and part of the lower extremities.
12. The SACRAL PLEXUS supplies the buttocks, perineum, and the lower extremities.
13. The VENTRAL RAMI of nerves T2-T11 do not form plexuses and are called INTERCOSTAL or THORACIC NERVES. They are distributed directly to the structures they supply in the intercostal spaces.

F. BRAIN: PRINCIPAL PARTS

1. The principal parts of the brain are the BRAIN STEM (medulla oblongata, pons, and midbrain), CEREBRUM, and CEREBELLUM.
2. The brain is PROTECTED by CRANIAL BONES, CRANIAL MENINGES, and CEREBROSPINAL FLUID.
3. The CRANIAL MENINGES are continuous with the spinal meninges and are named the DURA MATER, ARACHNOID, and PIA MATER.
4. The brain contains CAVITIES called VENTRICLES, which communicate with one another, with the CENTRAL CANAL of the spinal cord, and with the SUBARACHNOID SPACE.
5. CEREBROSPINAL FLUID is formed primarily by filtration from networks of capillaries called the CHOROID PLEXUS, found in the ventricles. It circulates through the ventricles, the central canal, and the subarachnoid space.
6. Most of the cerebrospinal fluid is absorbed by the arachnoid villi of the superior sagittal sinus, a blood sinus. The absorption occurs at the same rate the cerebrospinal fluid is produced, and thus a reasonably constant volume and pressure is maintained.
7. Cerebrospinal fluid protects the brain by serving as a shock absorber. It also delivers nutritive substances from the blood and removes waste materials.

G. CEREBROSPINAL FLUID

1. The brain and the CNS is protected from injury via the cerebrospinal fluid (CSF).
2. CSF circulates through the SUBARACHNOID SPACES around the brain and spinal cord, and through the VENTRICLES.
3. The VENTRICLES are cavities in the brain which are connected to one another, with the CENTRAL CANAL of the spinal cord and with the subarachnoid space.
4. There are four ventricles, two lateral ventricles, one third ventricle, and one forth ventricle.
5. The CNS contains 80-150 ml of CSF, which is a clear, colorless fluid composed of protein, glucose, urea, salts, and water.
6. The CSF functions in protection by acting as a shock absorbing medium for the brain and spinal cord.
7. The CSF functions in circulation by delivering nutritional substances in blood and removes wastes and toxic substances produced by the brain and spinal cord.
8. The CSF is formed by the filtration and secretion from the CHOROID PLEXUS, a specialized network of capillaries located within the ventricles.
9. The CSF flows in the following fashion. From the ventricles ---> to the subarachoid space around the brain and down around the posterior surface of the spinal cord --->up the anterior surface of the spinal cord --->around the anterior part of the brain.
10. From here the CSF is reabsorbed into veins called the SUPERIOR SAGITTAL SINUS through the ARACHNOID VILLI.

H. Blood Supply to the Brain

1. The brain composes only 2% of total body weight, but accounts for approximately 20% of the oxygen used by the body. It is the most metabolically active organ in the body, and the amount of oxygen it uses will vary with the degree of mental activity.
2. Any interruption of the oxygen supply to the brain can result in weakening, permanent damage, or death of brain cells.
3. Carbohydrate storage in the brain is severely limited, so the supply of glucose must be continuous. Glucose deficiency may produce mental confusion, dizziness, convulsions, and unconsciousness.
4. The BLOOD-BRAIN BARRIER explains the different rates of passage of certain material from the blood to the brain. The blood-brain barrier functions selectively to protect the brain from harmful substances.
5. The Circle of Willis supplies the entire brain with blood.

I. Medulla Oblongata

1. The MEDULLA OBLONGATA is continuous with the upper part of the spinal cord and contains portions of both motor and sensory tracts.
2. The RETICULAR FORMATION is a part of the medulla which functions in consciousness and arousal from sleep.
3. The medulla contains NUCLEI that serve as REFLEX CENTERS for the regulation of heart rate, respiratory rate, vasoconstriction, sneezing, coughing, hiccuping, and vomiting. The first three are considered vital reflexes.
4. The medulla also contains NUCLEI of ORIGIN for CRANIAL NERVES VIII and XII.

J. Pons

1. The PONS is located superior to the medulla. It connects the spinal cord with the brain and links one part of the brain with one another by way of tracts.
2. The PONS relays nerve impulses related to voluntary skeletal movements from the cortex of the cerebrum to the cerebellum.
3. It contains the NUCLEI for CRANIAL NERVES V-VII and a BRANCH OF CRANIAL NERVE VIII.
4. The RETICULAR FORMATION of the pons contains the PNEUMOTAXIC and APNEUSTIC centers which control respiration.

K. Midbrain

1. The MIDBRAIN, also known as the MESENCEPHALON, connects the pons and the diencephalon. It conveys motor impulses from the cerebrum to the cerebellum and spinal cord, sends sensory impulses from the spinal cord to the thalamus, and regulates auditory and visual reflexes.
2. It contains a major nucleus of the reticular formation, the red nucleus, and the nuclei of origin for cranial nerves III and IV.

L. Diencephalon

1. The DIENCEPHALON consists of the THALAMUS and HYOPOTHALAMUS.

2. The THALAMUS is located superior to the midbrain and contains nuclei that serve as relay stations for all sensory impulses except smell, traveling to the cerebral cortex. The THALAMUS also registers conscious recognition of pain and temperature, some awareness of light touch and pressure.
3. The HYPOTHALAMUS is inferior to the thalamus and contains important connections to the PITUITARY GLAND. It is capable of producing a variety of hormones that function in the maintenance of homeostasis.
4. The HYPOTHALAMUS CONTROLS and INTEGRATES the ANS, which regulates contraction of smooth muscle, cardiac muscle, and secretion of many glands. It is also involved in the reception and integration of sensory impulses from the body organs.
5. The HYPOTHALAMUS is the principal intermediary between the nervous system and the endocrine system and serves as a detector for many of the hormones in circulation.
6. It is associated with the feelings of aggression and rage and controls body temperature. It also functions to regulate food intake, fluid intake, and maintains the waking state and sleep patterns.
7. It is the center for biological rhythms. It also coordinates the mind-over-body phenomenon.

M. RETICULAR ACTIVATING SYSTEM (RAS), CONSCIOUSNESS, AND SLEEP

1. Humans awake and sleep in 24 hour rhythms called CIRCADIAN RHYTHMS.
2. The arousal state depends on the RETICULAR FORMATION which when stimulated increases cerebral activity. The area is also known as the RAS.
3. When the pons and the midbrain portions of the RAS are stimulated there is widespread cortical activity which results in consciousness and arousal.
4. Following arousal the RAS and cerebral cortex activate each other through a feedback system.
5. Inactivation of the RAS results in sleep or a partial state of unconsciousness.

N. CEREBRUM

1. The CEREBRUM is the largest part of the brain and contains a surface layer called the CEREBRAL CORTEX which is 2-4 millimeters thick and is comprised of GRAY MATTER.
2. The cortex contains GYRI OR CONVOLUTIONS, deep FISSURES, and shallower SULCI.
3. Beneath the cortex lies the cerebral white matter which has tracts which connect parts of the brain with itself and with other parts of the nervous system.
4. The cerebrum is nearly separated into RIGHT and LEFT HEMISPHERES by a CENTRAL LONGITUDINAL FISSURE. Internally, it remains connected by the CORPUS CALLOSUM, a bundle of transverse white fibers.

O. LOBES

1. Each cerebral hemisphere is subdivided into four LOBES by sulci or fissures.
2. The cerebral lobes are named the FRONTAL, PARIETAL, TEMPORAL, and OCCIPITAL.
3. A fifth part of the cerebrum, called the INSULA, lies deep to the parietal, frontal, and temporal lobes.

P. White Matter, Basal Ganglia, and Limbic System

1. The CEREBRAL WHITE MATTER, lying under the cortex, consists of myelinated axons running in three principal directions.
2. ASSOCIATION FIBERS connect and transmit nerve impulses between gyri in the same hemisphere; COMMISSURAL FIBERS connect gyri in one cerebral hemisphere with corresponding gyri in the other cerebral hemisphere; and PROJECTION FIBERS form ascending and descending tracts that transmit impulses from the cerebrum to other parts of the brain and spinal cord.
3. Paired masses of gray matter called BASAL GANGLIA lie within the cerebral hemispheres and help to control muscular movements.
4. The LIMBIC SYSTEM, which functions in emotional aspects of behavior and memory, lies within the cerebral hemispheres and the diencephalon.
5. The sensory areas of the cerebral cortex are concerned with the interpretation of sensory impulses. The motor areas of the cortex are the regions which govern muscular movement. The association areas of the cortex are concerned with emotional and intellectual processes.

Q. Brain Lateralization (Split-Brain Concept)

1. Recent research indicates that the two hemispheres of the brain are not bilaterally symmetrical, either anatomically or functionally.
2. The LEFT HEMISPHERE is more important for right-handed control, spoken and written language, numerical and scientific skills and reasoning.
3. The RIGHT HEMISPHERE is more important for left-handed control, musical and artistic awareness, space and pattern perception, insight, imagination, and generating mental images of sight, sound, taste, and smell.

R. Functional Areas of the Cerebral Cortex

1. The cerebral cortex is divided into the SENSORY AREAS, to receive and interpret sensory impulses, the MOTOR AREAS to control muscular movement, and ASSOCIATION AREAS which are concerned with integrative functions.
2. Integrative functions include memory, emotions, will, judgment, personality, and intelligence.
3. The principal sensory areas include the following:
   - PRIMARY SOMATOSENSORY AREA- posterior to the central sulcus
   - PRIMARY VISUAL AREA- occipital lobe; functions in shape and color
   - PRIMARY AUDITORY AREA- temporal lobe; functions in sound, pitch and rhythm
   - PRIMARY GUSTATORY AREA- base of the postcentral gyrus; functions in taste
   - PRIMARY OLFACTORY- temporal lobe; functions in smell
4. The principal motor areas are:
   - PRIMARY MOTOR AREA- precentral gyrus; functions in specific muscle movements
   - MOTOR SPEECH AREA (BROCA'S AREA)- left frontal lobe; functions in speech
5. The principal association areas are as follows:
   - SOMATOSENSORY ASSOCIATION AREA- posterior to the primary somatosensory area; functions to interpret and integrate sensations
   - VISUAL ASSOCIATION AREA- occipital lobe; functions to relate past and present visual experiences
   - AUDITORY ASSOCIATION AREA- temporal lobe; functions to determine if sound is speech, music, or noise, also functions to translate words into thoughts

- GNOSTIC AREA- integrates visual and auditory association areas to form a common thought from a stimulus
- PREMOTOR AREA- anterior to the primary motor area; deals with functions which are of a complex and sequential nature
- FRONTAL EYE FIELD AREA- frontal lobe; functions in eye scanning
- LANGUAGE AREA- incorporates all lobes

S. CEREBELLUM

1. The CEREBELLUM occupies the inferior and posterior aspects of the cranial cavity. It consists of two HEMISPHERES and a CENTRAL, CONSTRICTED VERMIS.
2. It is attached to the brainstem by three pairs of CEREBRAL PEDUCLES and functions in the coordination of skeletal muscles and the maintenance of normal muscle tone and body equilibrium.

T. NEUROTRANSMITTERS IN THE BRAIN

1. There are numerous substances that are either known or suspected neurotransmitters in the brain. These substances can facilitate, excite, or inhibit postsynaptic neurons.
2. Examples of neurotransmitters include acetylcholine, glutamic acid, aspartic acid, norepinephrine, dopamine, seratonin, gamma aminobutyric acid and glycine.
3. Neuropeptides are another group of chemical messengers that have been identified in the brain. Many of these act as natural pain killers and include such substances as enkephalins, endorphins, substance P and dynorphin.
4. Other neuropeptides serve as hormones or regulators of other physiological responses. These may include angiotensin, cholecystikinin, and regulating hormones produced by the hypothalamus.

U. CRANIAL NERVES

1. Twelve pairs of cranial nerves originate from the brain.
2. The pairs are named primarily on the basis of distribution, and are numbered by order of attachment to the brain by roman numerals.
3. Some cranial nerves I, II and VIII contain only sensory fibers, and are called sensory nerves. The others are mixed nerves because they contain both sensory and motor fibers.

CRANIAL NERVES

NUMBER	NERVE	TYPE	FUNCTION
I	OLFACTORY	SENSORY	SMELL
II	OPTIC	SENSORY	VISION
III	OCULOMOTOR	MIXED, PRIMARILY MOTOR	MOTOR FUNCTION: MOVEMENT OF EYELID AND EYEBALL, ACCOMMODATION OF LENS FOR NEAR VISION AND CONSTRICTION OF THE PUPIL; SENSORY FUNCTION: PROPRIOCEPTION
IV	TROCHLEAR	MIXED, PRIMARILY MOTOR	MOTOR FUNCTION: MOVEMENT OF THE EYEBALL; SENSORY FUNCTION: MUSCLE SENSE
V	TRIGEMINAL	MIXED	MOTOR FUNCTION: CHEWING; SENSORY FUNCTION; CONVEYS SENSATIONS FOR TOUCH, PAIN AND TEMPERATURE FROM STRUCTURES SUPPLIED ; SENSORY FUNCTION: PROPRIOCEPTION
VI	ABDUCENS	MIXED, PRIMARILY MOTOR	MOTOR FUNCTION: MOVEMENT OF THE EYEBALL; SENSORY FUNCTION: PROPRIOCEPTION
VII	FACIAL	MIXED	MOTOR FUNCTION: FACIAL EXPRESSION AND SECRETION OF SALIVA AND TEARS; SENSORY FUNCTION: PROPRIOCEPTION
VIII	VESTIBULO-COCHLEAR	SENSORY	HEARING AND EQUILIBRIUM
IX	GLOSSO-PHARYNGEAL	MIXED	MOTOR FUNCTION: SECRETION OF SALIVA; SENSORY FUNCTION: TASTE AND REGULATION OF BLOOD PRESSURE, PROPRIOCEPTION
X	VAGUS	MIXED	MOTOR FUNCTION: VISCERAL MUSCLE MOVEMENT; SENSORY FUNCTION: SENSATIONS FROM ORGANS SUPPLIED, PROPRIOCEPTION
XI	ACCESSORY	MIXED, PRIMARILY MOTOR	MOTOR FUNCTION: MEDIATES SWALLOWING MOVEMENTS AND MOVEMENTS OF THE HEAD; SENSORY FUNCTION: PROPRIOCEPTION
XII	HYPOGLOSSAL	MIXED, PRIMARILY MOTOR	MOTOR FUNCTION: MOVEMENT OF THE TONGUE DURING SPEECH AND SWALLOWING; SENSORY FUNCTION: PROPRIOCEPTION

IV. TEACHING TIPS AND SUGGESTIONS

A. HELPFUL HINTS

1. When detailing the anatomy of spinal nerves, it would be useful to use a model as well as a complete vertebral column.
2. When detailing the plexuses, it is very useful to have a dissection specimen on hand, since diagrams cannot convey the three-dimensional nature of plexuses.
3. Dissection of a sheep or comparable brain is useful in demonstrating the lobes and anatomy, as well as to illustrate the origins of the cranial nerves.

B. ESSAY QUESTIONS

1. After touching a hot stove you immediately withdraw your hand and step back. What kind of reflex arc is involved? Where are the receptors located and what is their function? Where is the center of this reflex arc? What is the role of the effectors?
2. Explain, after examination of the spinal nerves, why spinal segment 25 does not lie under vertebra 25. Why would this knowledge be useful in performing a spinal or lumbar puncture?
3. Trace the flow of cerebrospinal fluid from where it originates in the choroid plexus to where it is reabsorbed in the veins. Name all of the structures through which it circulates. How many ventricles are involved?

C. TOPIC FOR DISCUSSION

1. Discuss the role of the hypothalamus in feeding and satiety under normal conditions. Elaborate by introducing the concept that this center can be overridden by voluntary impulses arising in the cerebral cortex. Emphasize how our perception of how much we eat often exceeds our body's own judgement.

V. AUDIOVISUAL MATERIALS

A. VIDEOCASSETTES

1. The Talented Brain (26 min.; FHS).
2. Decision (26 min.; FHS).
3. The Sexual Brain (28 min.; FHS).
4. Depression (30 min.; PLP).
5. The Enlightened Machine (60 min.; 1984; FI/KSU).
6. Rhythms and Drives (60 min.; 1984; KSU).
7. States of Mind (60 min.; 1984; KSU).
8. The Two Brains (60 min.; 1984; KSU).
9. Stress and Emotion (60 min; 1984; KSU).
10. The Nervous System: Nerves At Work (26 min.; 1985; FHS).
11. The Addicted Brain (24 min.; 1986; FHS).

B. FILMS: 16 MM

1. The Human Body: The Brain (16 min.; 1968; COR/KSU).
2. Exploring the Human Brain (19 min.; 1977; KSU).
3. Mind Over Body (49 min.; 1973; TLF).
4. The Human Brain (24 min.; 1983; EBEC/KSU).
5. How the Mind Begins (24 min.; TGC).
6. The Mind of Man (119 min.; UIFC).
7. Brain and Behavior (22 min.; 1957; UIFC/McG/KSU).
8. Marvel of the Brain (25 min.; NGF).
9. The Hidden Universe: The Brain (48 min.; 1977; McG).
10. Secrets of the Brain (15 min.; FHS).

C. TRANSPARENCIES: 35 MM (2X2)

1. PAP Slide Set (Slides 55-59).
2. AHA Slide Set.
3. Neurotransmitters (75 Slides; IBIS).
4. Psychobiology: The Brain and Behavior (150 Slides; CARO).

D. OVERHEAD TRANSPARENCIES

1. PAP Transparency Set (Trs. 14.1a, 14.2, 14.3a, 14.4a, 14.5a&b, 14.6, 14.7, 14.8a&b, 14.9, 14.10a&c, 14.12, 14.13, 14.14, 14.17a&b, 14.18).
2. Nervous System: Brain (GAF).
3. Nervous System (HSC).
4. Nervous System- Unit 5 (RJB).
5. Human Brain Anatomy, Parts 1-2 (2 Transparencies; CARO).

E. COMPUTER SOFTWARE

1. Nervous System (Apple; SC-182013; IBM; SC-182014; PLP).
2. Brain Probe (Apple; SC-175034; PLP).
3. The Human Brain: Neurons (Apple; SC-510003; PLP).
4. Body Language: The Nervous System (Apple II Series; ESF).
5. Cocaine (Apple II; IBM-PC; CARO).
6. Drugs: Body & Mind/Cocaine (Apple; IBM; MAC; PLP).
7. Drugs: Body & Mind/Drinking and Not Drinking (Apple; IBM; MAC; PLP).
8. Drugs: Body & Mind/Introduction to Psychoactive Drugs (Apple; IBM; MAC; PLP).
9. Drugs: Body & Mind/Keeping Off the Grass-Marijuana (Apple; IBM; MAC; PLP).
10. Drugs: Body & Mind/Six Classes of Psychoactive Drugs (Apple; IBM; MAC; PLP).
11. Drugs: Body & Mind/Tobacco (Apple; IBM; MAC; PLP).
12. Flash: Nerves (Apple; IBM; PLP).
13. Flash: EEG (Apple; IBM; PLP).
14. Flash: The Human Brain (Apple; IBM; PLP).

CHAPTER 11

AUTONOMIC NERVOUS SYSTEM

CHAPTER AT A GLANCE

- COMPARISON OF SOMATIC AND AUTONOMIC NERVOUS SYSTEMS
- STRUCTURE OF THE AUTONOMIC NERVOUS SYSTEM
- *Autonomic Motor Pathways*
- *Preganglionic Neurons*
- *Autonomic Ganglia*
- *Postganglionic Neurons*
- *Sympathetic Division*
- *Parasympathetic Division*
- FUNCTIONS OF THE AUTONOMIC NERVOUS SYSTEM (ANS)
- *ANS Neurotransmitters*
- *Activities*
- WELLNESS FOCUS: STRESS RESISTANCE - IT DEPENDS ON YOUR POINT OF VIEW

I. CHAPTER SYNOPSIS

Students are introduced to the general functions of the ANS and its comparison to the SNS. The sympathetic and parasympathetic divisions of the autonomic nervous system are introduced. Attention is then focused on the structure of the ANS, detailing the components of visceral efferent pathways. The components discussed are preganglionic neurons, autonomic ganglia, and postganglionic neurons. The functional aspects of the autonomic nervous system are then discussed by considering neurotransmitters and numerous ANS activities. The chapter concludes with the effects of sympathetic and parasympathetic divisions on glands, muscles, and the visceral components of the body.

II. LEARNING GOALS/STUDENT OBJECTIVES

1. Compare the main structural and functional differences between the somatic and autonomic nervous systems.
2. Identify the structural features of the autonomic nervous system.
3. Describe the functions of the autonomic nervous system.

III. SAMPLE LECTURE OUTLINE

A. INTRODUCTION

1. The ANS regulates the activities of smooth muscle, cardiac muscle, and glands. It co-functions with the SNS to form the PNS.
2. The ANS is comprised of VISCERAL EFFERENT NEURONS which are organized into nerves, ganglia, and plexuses, which generally operate without conscious control.
3. The autonomic centers located in the cerebrum cortex are connected to the autonomic centers of the thalamus and the hypothalamus. The hypothalamus exerts the majority of control over the ANS.

B. COMPARISON OF AUTONOMIC AND SOMATIC NERVOUS SYSTEMS

1. In contrast to the SNS, which conveys both SENSORY and CONSCIOUS MOTOR INFORMATION, the ANS regulates VISCERAL ACTIVITIES INVOLUNTARILY and AUTONOMICALLY.
2. AFFERENT (SENSORY) IMPULSES from visceral effectors travel to the autonomic centers of the brain along afferent neurons.
3. VISCERAL EFFERENT (MOTOR) FIBERS transmit impulses from the central nervous system to visceral effectors.
4. Efferent impulses usually produce an adjustment in a visceral effector without conscious recognition. Exceptions include hunger, nausea, and fullness of the urinary bladder.
5. The two subdivisions of the ANS are the SYMPATHETIC and PARASYMPATHETIC divisions. Most glands are innervated by visceral effector neurons from both divisions.
6. The two subdivisions are generally ANTAGONISTIC. If one subdivision increases a specific activity, the other subdivision will function to decrease that activity at the appropriate time.
7. Organs receiving both sympathetic and parasympathetic innervation are said to have DUAL INNERVATION. This innervation can therefore be either excitatory or inhibitory.
8. The ANS contains two EFFERENT NEURONS and a GANGLION positioned between the two neurons. The first neuron transmits information from the CNS to the ganglion where it synapses with the second neuron. The second neuron will synapse on a visceral effector.
9. The SNS releases ACH while the ANS releases both ACH and NOREPINEPHRINE (NE).

C. STRUCTURE OF THE ANS

1. Autonomic motor pathways involve two efferent neurons: the PREGANGLIONIC FIBER, which extends from the central nervous system to the ganglion, and the POSTGANGLIONIC FIBER, which extends from the ganglion to the effector.
2. The PREGANGLIONIC NEURON has its cell body in the CNS. It is myelinated and passes out of the the CNS as part of the spinal nerve.
3. The POSTGANGLIONIC NEURON has its cell body outside of the CNS. It is unmyelinated and terminates at a visceral effector.

D. PREGANGLIONIC NEURONS

1. In the SYMPATHETIC DIVISION, PREGANGLIONIC NEURONS have their cell bodies in the lateral gray horns of the twelve thoracic segments and first and second lumbar segments of the spinal cord.

2. In the PARASYMPATHETIC DIVISION, the cell bodies are located in the nuclei of cranial nerves III, VII, IX, and X in the brain stem, and in the lateral gray horns of the second, third, and fourth sacral segments of the spinal cord.

E. AUTONOMIC GANGLION

1. All AUTONOMIC PATHWAYS include autonomic ganglia, a region where the visceral effector neurons synapse.
2. The THREE GROUPS of autonomic ganglia are the SYMPATHETIC TRUNK GANGLION, PREVERTEBRAL GANGLION, and the TERMINAL GANGLION.
3. The SYMPATHETIC TRUNK GANGLIA lie on either side of the vertebral column from the base of the skull to the coccyx and receive preganglionic fibers from the sympathetic division. The sympathetic preganglionic fibers are characteristically short.
4. The PREVERTEBRAL GANGLION lies outside the spinal column and near the large abdominal arteries from which the prevertebral ganglia obtain their names.
5. The PREVERTEBRAL GANGLION are named for the proximate abdominal arteries and are called CELIAC GANGLION, SUPERIOR MESENTERIC GANGLION and INFERIOR MESENTERIC GANGLION. They all receive preganglionic fibers from the sympathetic division.
6. The TERMINAL GANGLIA belong to the parasympathetic division and are located at the end of visceral effector pathways close to or within the walls of the effectors. These preganglionic fibers are characteristically long.

F. POSTGANGLIONIC NEURONS

1. POSTGANGLIONIC NEURONS exit the ganglia and either innervate or supply visceral effectors. If the preganglionic fiber in the pathway is long, the postganglionic fiber will be short. If the preganglionic fiber in the pathway is short, the postganglionic fiber will be long.

G. SYMPATHETIC DIVISION

1. Preganglionic fibers are myelinated and leave the spinal cord by way of the anterior root.
2. After exiting the spinal cord they go to the sympathetic trunk ganglion on the same side.
3. A preganglionic fiber may terminate with a large number of postganglionic cell bodies in the ganglion.
4. An impulse originating in a single preganglionic fiber will terminate in a visceral effector.

H. PARASYMPATHETIC DIVISION

1. Preganglionic cell bodies of the parasympathetic division are found in nuclei in the brainstem and sacral segments of the spinal cord.
2. The fibers emerge as part of a cranial nerve or spinal nerve.
3. Preganglionic fibers of both cranial and sacral outflows end in terminal glanglia where they synapse with postganglionic neurons.
4. The synapse is local and precise.

I. Functions of the ANS

1. Autonomic fibers release neurotransmitters at synapses and at contact points with visceral effectors. The visceral effector contact points are called neuroeffector junctions and are either neuromuscular or neuroglandular junctions.
2. Autonomic fibers are classified as being either cholinergic or adrenergic, depending on the neurotransmitter released.
3. Cholinergic fibers release AcH and include all sympathetic and parasympathetic preganglionic axons, all postganglionic axons, and a select group of sympathetic postganglionic axons which innervate sweat glands and blood vessels in skeletal muscle.
4. Adrenergic fibers release NE as a neurotransmitter, which is characteristically released by sympathetic postganglionic fibers.

J. ANS Activities

1. The sympathetic and parasympathetic divisions are antagonistic toward a particular effector. The stimulatory division may be either sympathetic or parasympathetic, depending upon the organ. For example, the sympathetic division increases heart rate and inhibits digestive processes, whereas the parasympathetic division enhances digestive processes and slows down heart rate.
2. The action of the two divisions on a visceral effector help to maintain and sustain dynamic homeostasis.
3. The parasympathetic division is principally concerned with activities that conserve energy where the sympathetic division is concerned with activities that expend energy.
4. Under stressful situations, the sympathetic division dominates and evokes a physiological response called the fight-or-flight response.

IV. TEACHING TIPS AND SUGGESTIONS

A. Helpful Hints

1. It would be useful to itemize the major function of the SNS, and to explain how it relays both sensory and motor information.
2. A cross-section of the spinal cord can be used to show the anterior and posterior roots and how both sensory and motor fibers can be relayed together. Contrast this to how the autonomic fibers emanate from the spinal cord.
3. Explain the differences between somatic reflexes and visceral reflexes.

B. Essay Questions

1. How are the sympathetic and parasympathetic divisions of the autonomic nervous system differentiated structurally and functionally?
2. Is there a functional reason for the existence of short and long pre- and postganglionic fibers?

C. Topic for Discussion

1. Discuss the concept of biofeedback in the voluntary regulation of normally involuntary actions such as heart rate and breathing.

V. AUDIOVISUAL MATERIALS

A. Films: 16 mm

1. Parasympathetic and Sympathetic Innervation, Part I-II (40 min.; FHS).
2. Mind Over Body (49 min.; 1973; IFB).
3. Autonomic Nervous System (17 min.; 1975; IFB).
4. The Autonomic Nervous System-An Overview (17 min.; 1973; USNAC).

B. Transparencies: 35 mm (2x2)

1. PAP Slide Set (slides 65-67).
2. AHA Slide Set.

C. Overhead Transparencies

1. The Autonomic Nervous System (GAF).
2. The Autonomic Nervous System (1 transparency with 3 overlays, CARO).

CHAPTER 12

SENSATIONS

CHAPTER AT A GLANCE

- SENSATIONS
  - *Definition*
  - *Characteristics*
  - *Classification of Receptors*
- GENERAL SENSES
  - *Cutaneous Sensations*
    - *Tactile Sensations*
    - *Thermal Sensations*
    - *Pain Sensations*
  - *Proprioceptive Sensations*
    - *Receptors*
- SPECIAL SENSES
- OLFACTORY SENSATIONS
  - *Structure of Receptors*
  - *Stimulation of Receptors*
  - *Olfactory Pathway*
- GUSTATORY SENSATIONS
  - *Structure of Receptors*
  - *Stimulation of Receptors*
  - *Gustatory Pathway*
- VISUAL SENSATIONS
- ACCESSORY STRUCTURES OF THE EYE
  - *Structure of the Eyeball*
    - *Fibrous Tunic*
    - *Vascular Tunic*
    - *Retina (Nervous Tunic)*
    - *Lens*
    - *Interior*
  - *Physiology of Vision*
    - *Retinal Image Formation*
    - *Stimulation of Photoreceptors*
    - *Visual Pathway*
- AUDITORY SENSATIONS AND EQUILIBRIUM
  - *External Ear*
  - *Middle Ear*
  - *Internal Ear*
  - *Sound Waves*
  - *Physiology of Hearing*
  - *Physiology of Equilibrium*
    - *Static Equilibrium*

- *Dynamic Equilibrium*
- COMMON DISORDERS
- MEDICAL TERMINOLOGY AND CONDITIONS
- WELLNESS FOCUS: THE EYES (AND EARS) HAVE IT

I. CHAPTER SYNOPSIS

This chapter offers a general introduction to the location of sensory apparati and the way in which stimuli are detected. General cutaneous sensations such as tactile and discriminative touch, pressure, pain, and temperature are considered. The special senses of smell, taste, sight, and hearing are then introduced with a discussion of the anatomical and physiological features. The chapter concludes with a list of common disorders, medical terms, and conditions associated with the sense organs.

II. LEARNING GOALS/STUDENT OBJECTIVES

1. Define a sensation and describe the conditions necessary for a sensation to occur.
2. List and describe the cutaneous sensations.
3. Define proprioception and describe the structure of proprioceptive receptors.
4. Describe the receptors for olfaction and the olfactory pathway to the brain.
5. Describe the receptors for gustation and the gustatory pathway to the brain.
6. Describe the receptors for visual sensations and the visual pathways to the brain.
7. Describe the mechanism involved in vision.
8. Describe the receptors for hearing and equilibrium, and their pathways to the brain.

III. SAMPLE LECTURE OUTLINE

A. CHARACTERISTICS AND CLASSIFICATION OF SENSATIONS

1. A SENSATION is a state of awareness of external or internal conditions of the body. Four conditions must be satisfied for a sensation to occur. There must be a STIMULUS, a RECEPTOR, or sense organ to receive the stimulus, a NERVE PATHWAY to conduct the information to the brain, and a region of the brain to TRANSLATE the sensation.
2. Stimuli create changes in membrane potentials at receptors, resulting in a change in the cell's resting potential. This is called GENERATOR POTENTIAL.
3. Conscious sensations and perceptions occur in the cerebral cortical regions of the brain. The brain uses the processes of projection, adaptation, afterimaging, and modality.
4. PROJECTION is the process by which the brain refers sensations to the point of origin.
5. ADAPTATION is the process in which the brain decreases its sensitivity to a stimulus when the stimulus is continuously applied.
6. AFTERIMAGING is the process whereby the sensation persists even after the stimulus has been removed.
7. MODALITY refers to that specific characteristic of each sensation which allows it to be distinguished from other types.

8. Receptors can be classified as EXTERORECEPTORS, which provide information about the external environment; ENTERORECEPTORS, which provide information about the internal environment; and PROPRIOCEPTORS, which provide information concerning body position and movement.
9. The stimuli detected are designated by specific terms. MECHANORECEPTORS, detect mechanical deformation of adjacent cells; THERMORECEPTORS, detect changes in temperature; NOCICEPTORS, detect pain; PHOTORECEPTORS, detect light; and CHEMORECEPTORS detect the presence of chemicals in solution.

B. CUTANEOUS SENSATIONS

1. CUTANEOUS SENSATIONS include TACTILE SENSATIONS (touch, pressure, & vibration), THERMORECEPTIVE SENSATIONS, and PAIN.
2. Receptors are widely, and unevenly distributed over the body so that some areas are more sensitive to stimuli than others.
3. All perception and interpretation of cutaneous sensation is accomplished in the PARIETAL LOBE of the cerebral cortex.
4. All TACTILE SENSATIONS are detected by MECHANORECEPTORS.
5. Touch sensations arise from tactile receptors beneath the skin. Light touch indicates sensations in a general area while discriminative touch is perceived as a specific and identifiable point on the body.
6. Tactile receptors include hair root plexuses, free nerve endings, tactile discs, corpuscles of touch, and type II cutaneous mechanoreceptors.
7. Hair root plexuses detect movements when the hair shaft is moved or disturbed.
8. Free nerve endings are found everywhere in the skin and are important in pain reception and the reception of continuous body contact.
9. TACTILE, or MERKEL'S DISCS, are found in hairless skin and connect to free nerve endings.
10. CORPUSCLES OF TOUCH, or MEISSNER'S CORPUSCLES, are located in the fingertips, dermal papillae of the skin, palms, soles, eyelids, tongue, nipples, clitoris, and the tip of the penis. They function in discriminative touch.
11. TYPE II CUTANEOUS MECHANORECEPTORS, or ORGANS OF RUFFINI, detect heavy and continuous touch and are embedded in the deep dermis.
12. PRESSURE RECEPTION is accomplished by the PACINIAN CORPUSCLES located in the subcutaneous layer of the skin, around joints, tendons and muscles, in the mammary glands, external genitalia, and some viscera.
13. The mechanics of thermoreception are not known. Temperatures as low as 50°F and as high as 113°F can be detected.
14. The receptors for pain are called nociceptors and are free nerve endings. They are found in all tissues of the body and respond to any stimulus.

C. PROPRIOCEPTIVE SENSATIONS

1. PROPRIOCEPTORS provide information concerning movement. These receptors are located in the skeletal muscles, tendons, around synovial joints, and in the internal ear.
2. Proprioception is also known as the KINESTHETIC SENSE.
3. MUSCLE SPINDLES are delicate proprioceptive receptors found between skeletal muscle fibers. When a muscle is stretched, the receptors are stimulated and relay information to the CNS indicating the activity of the muscle.

4. TENDON ORGANS are found at the junction of muscle and tendon. They protect the tendon and associated muscle from damage due to excessive tension.
5. JOINT KINETIC RECEPTORS are located around the synovial joints and respond to pressure, acceleration, deceleration, and excessive strain placed on the joint.
6. MACULAE and CRISTAE are located in the inner ear and function in equilibrium.

D. OLFACTORY SENSATIONS

1. The OLFACTORY CELLS, the receptors for olfaction, are located in the nasal epithelium in the superior portion of the nasal cavity.
2. In order to be smelled, substances must be volatile, water-soluble, and lipid-soluble.
3. The chemical theory assumes that there are different receptor molecules in the membranes of the olfactory hairs, each capable of reacting with a particular stimulus. The interaction of the olfactory cells and the substance alters the permeability of the plasma membrane so that a generator potential is established. This is followed by the initiation of a nerve impulse.
4. Adaptation to odors occurs quickly, and the threshold of smell is low.
5. Olfactory cells convey nerve impulses to the olfactory nerves (cranial nerve I), olfactory bulbs, tracts, and the cerebral cortex.

E. GUSTATORY SENSATIONS

1. The GUSTATORY CELLS, the receptors for TASTE, are located in taste buds on the surface of the tongue.
2. Substances to be tasted must be in solution in saliva.
3. The four primary tastes are SALT, SWEET, BITTER, and SOUR.
4. The senses of smell and taste are closely associated. If one's ability to smell is impaired, the ability to taste will be greatly diminished.
5. Adaptation to taste occurs quickly and the threshold for taste varies for each of the primary tastes.
6. Gustatory cells convey their impulses to the facial (VII), glossopharyngeal (IX), vagus (X) cranial nerves, medulla, thalamus, and the cerebral cortex.

F. VISUAL SENSATIONS

1. ACCESSORY STRUCTURES of the eyes include EYEBROWS, EYELIDS, EYELASHES, and the LACRIMAL APPARATUS. The CONJUNCTIVA is a thin mucous membrane which lines the inner aspect of the eyelids and is reflected onto the anterior surface of the eyeball.
2. The EYE is constructed of THREE LAYERS: the OUTER FIBROUS TUNIC, THE MIDDLE VASCULAR TUNIC, and the INNER RETINA.
3. The FIBROUS TUNIC is divided into two regions: the posterior SCLERA and the anterior CORNEA.
4. The SCLERA is also known as the white of the eye and is comprised of dense fibrous connective tissue. It covers the entire eye except for the most anterior portion. It provides shape and affords protection to the inner parts. Its posterior aspect is pierced by the OPTIC NERVE (cranial nerve).
5. The CORNEA is a nonvascular, transparent, fibrous coat through which the iris can be seen. The cornea acts in the refraction of light.
6. The VASCULAR TUNIC, or middle layer, is composed of three portions: the CHOROID, CILIARY BODY and IRIS.

7. The CHOROID absorbs light rays so that they are not reflected within the eyeball, and nourishes the retina through its blood supply.
8. The CILIARY BODY consists of the ciliary processes and ciliary muscles.
9. The CILIARY PROCESSES consist of protrusions or folds on the internal surface of the ciliary body that secrete aqueous humor.
10. The CILIARY MUSCLES are smooth muscle that alter the shape of the lens for near and far vision.
11. The IRIS is the circular colored portion seen through the cornea. It consists of smooth muscle fibers and together make a donut-shaped structure.
12. The black hole in the center of the iris is the PUPIL, the area through which light enters the eyeball. The IRIS functions to regulate the amount of light entering the posterior cavity of the eye.
13. The RETINA, or inner layer, lies in the posterior portion of the eye and functions in image formation.
14. The NERVOUS LAYER of the RETINA contains three zones of neurons which are named in the order they conduct nerve impulses. These are: PHOTORECEPTOR NEURONS, BIPOLAR NEURONS, and GANGLION NEURONS.
15. The dendrites of the photoreceptor neurons are called RODS and CONES because of their shape. The rods are instrumental in BLACK and WHITE vision in dim light; the CONES are instrumental in color vision and in visual acuity (color vision in bright light).
16. The CONES are densely concentrated in a small depression in the posterior portion of the eye called the CENTRAL FOVEA. The central fovea is located in the MACULA LUTEA, which is the exact center of the posterior portion of the eye. The macula lutea corresponds to the visual axis of the eye. The central fovea is the area of the sharpest vision because of the high concentration of cones.
17. Posterior to the pupil and iris is the lens which aids to focus light rays for clear vision.
18. The interior of the eye is divided into the ANTERIOR CAVITY and the POSTERIOR CAVITY. The ANTERIOR CAVITY is filled with a watery fluid which is called the AQUEOUS HUMOR. The aqueous humor is continuously produced. The POSTERIOR CAVITY is the largest of the cavities and lies between the retina and the lens. It contains a non-replenished, jelly-like substance called VITREOUS HUMOR.

G. PHYSIOLOGY OF VISION

1. RETINAL IMAGE FORMATION involves the refraction of light, accommodation of the lens, constriction of the pupil, convergence, and inverted image formation.
2. REFRACTION is the bending of light rays where two different media meet. The refraction media are the cornea, aqueous humor, lens, and vitreous humor.
3. ACCOMMODATION is the ability of the lens to instantly change its curvature to allow for changes in focus from a near object to one that is far away. The ciliary muscle contracts for near objects and relaxes for far objects.
4. Constriction of the pupil occurs simultaneously with lens accommodation and inhibits light rays from entering the eye through the periphery of the lens.
5. CONVERGENCE is the medial movement of both eyeballs so that they are directed on the object being viewed. This allows for binocular vision.
6. Images are focused on the retina upside down and undergo a mirror image reversal. These images are then rearranged by the brain to produce an image perceived in the actual orientation.

7. The rods and cones develop receptor potentials and the ganglion cells initiate the nerve impulse. The rods contain a photopigment called rhodopsin that undergoes structural changes and leads to the development of a receptor potential.
8. Impulses from ganglion cells are conveyed through the retina to the OPTIC NERVE (CRANIAL NERVE II), the OPTIC CHIASMA, OPTIC TRACT, the THALAMUS, and the CEREBRAL CORTEX.
9. Sight is perceived in the OCCIPITAL LOBE of the brain.

H. AUDITORY SENSATIONS

1. The EAR consists of three anatomical subdivisions: the EXTERNAL EAR, MIDDLE EAR, and the INNER EAR.
2. The OUTER EAR is constructed to direct sound waves. It is comprised of the PINNA, EXTERNAL AUDITORY CANAL, and the TYMPANIC MEMBRANE.
3. The MIDDLE EAR, or TYMPANIC CAVITY, is a hollow cavity lined with epithelium. It maintains continuity with the nasopharynx by way of the Eustachian tubes.
4. The middle ear is hollowed out of the TEMPORAL BONE and is separated from the outer ear by the EARDRUM and the inner ear by a thin bony partition with two small openings: the OVAL WINDOW and the ROUND WINDOW.
5. The INNER EAR, or LABYRINTH, is comprised of a series of canals. It consists of an OUTER BONY LABYRINTH and an INNER MEMBRANOUS LABYRINTH that fits into the bony labyrinth.
6. The BONY LABYRINTH is part of the petrous portion of the temporal bone and is divided by shape into the VESTIBULE, COCHLEA, and SEMICIRCULAR CANALS.
7. The bony labyrinth is LINED WITH PERIOSTEUM and contains a fluid called PERILYMPH, which surrounds the membranous labyrinth.
8. The MEMBRANOUS LABYRINTH is lined with epithelium and contains a fluid called ENDOLYMPH.
9. The VESTIBULE constitutes the oval portion of the bony labyrinth. The membranous labyrinth in the vestibule consists of two sacs called the UTRICLE and SACCULE.
10. Projecting superiorly and posteriorly from the vestibule are three SEMICIRCULAR CANALS arranged at right angles to one another.
11. Lying anterior to the vestibule is the COCHLEA, a spiral bony canal which is divided into three regions; the SCALA VESTIBULI, which ends at the oval window, the SCALA TYMPANI which terminates at the round window, and the SCALA MEDIA, which is separated from the SCALA TYMPANI by the BASILAR MEMBRANE.
12. The ORGAN of CORTI is the principal organ of hearing. It rests on the BASILAR MEMBRANE.
13. Projecting over and in contact with the ORGAN of CORTI are HAIR CELLS, which are part of the TECTORIAL MEMBRANE.

I. PHYSIOLOGY OF HEARING

1. SOUND WAVES enter the ear through the EXTERNAL AUDITORY MEATUS, strike the TYMPANIC MEMBRANE, and are conducted through the OSSICLES, the MALLEUS, INCUS and the STAPES.
2. The waves strike the OVAL WINDOW, which sets up waves in the PERILYMPH of the SCALA VESTIBULI.
3. As the pressure moves through the perilymph of the scala vestibuli, the waves of the fluid strike the VESTIBULAR MEMBRANE and SCALA TYMPANI.
4. The vestibular membrane is pushed inward and increases the pressure of the ENDOLYMPH in the duct of the COCHLEA.

5. The pressure waves strike the BASILAR MEMBRANE, stimulation of the hairs on the ORGAN of CORTI occurs, and a nerve impulse is initiated.
6. Nerve impulses from the COCHLEAR BRANCH of the VESTIBULOCOCHLEAR NERVE (VIII) pass to the COCHLEAR NUCLEI of the MEDULLA. Here, they cross to opposite sides and travel to the MIDBRAIN, THALAMUS, and finally to the AUDITORY REGION of the TEMPORAL LOBE of the cerebral cortex.

J. PHYSIOLOGY OF EQUILIBRIUM

1. There are two kinds of equilibrium.
2. STATIC EQUILIBRIUM refers to the position of the body, principally the head, relative to the ground.
3. DYNAMIC EQUILIBRIUM is the maintenance of the body position in response to sudden movements.
4. The receptors organs for equilibrium are in the internal ear. They are the SACCULE, UTRICLE, and the SEMICIRCULAR DUCTS.

K. Static Equilibrium

1. The walls of the UTRICLE and SACCULE contain a small, flat region called the MACULA.
2. The MACULAE are the receptors for static equilibrium and consist of two kinds of cells, the HAIR (RECEPTOR) CELLS and SUPPORTING CELLS.
3. The HAIR CELLS contain long extensions of the cell membrane consisting of many STEREOCILIA and one extremely long KINOCILIUM.
4. Floating over the hair cells is a jelly-like material called OTOLITHIC MEMBRANE.
5. The OTOLITHIC MEMBRANE sits on top of the macula and move relative to the position of your head.
6. Movement pulls on the STEREOCILIA and makes them bend. This initiates a nerve impulse that is then transmitted to the vestibular branch of the vestibulocochlear nerve (VIII).

L. DYNAMIC EQUILIBRIUM

1. The three semicircular ducts maintain dynamic equilibrium.
2. The ducts are positioned at right angles to one another in three planes.
3. This positioning permits detection of an imbalance in three planes.
4. In the AMPULLA, the dilated portion of each duct, there is a small elevation called the CRISTA.
5. Each crista is composed of a group of HAIR RECEPTOR CELLS and SUPPORTING CELLS covered by a jelly-like material called the CUPULA.
6. When the head moves, the ENDOLYMPH in the semicircular ducts flows over the air cells and bends them.
7. Movement of the hair cells stimulates sensory neurons and cranial nerve VIII.

IV. TEACHING TIPS AND SUGGESTIONS

A. Helpful Hints

1. Taste sensations can be demonstrated in the lab by having the students sprinkle salt or sugar onto a dry tongue. Without saliva to dissolve the substance, the sensation of taste is markedly reduced.
2. The use of an opthalmoscope to examine the interior of the eye is a useful laboratory aid.
3. Dynamic and static equilibrium can be demonstrated by asking a volunteer to allow himself to be spun in a chair with a bag over his head for 20 seconds, then without the bag for an additional 20 seconds.

B. Essay Questions

1. The lacrimal glands release lacrimal fluid onto the eyes. Explain why this fluid is important and the role of the conjunctival membranes and eyelids in the dispersement of the lacrimal fluid.
2. Describe the origin and pathway of an impulse that results in smell.
3. There are two individuals in class: one with myopia and the other with hypermetropia. What are these conditions and how can they be corrected? What is the cause for each of these conditions?

C. Topic for Discussion

1. Discuss the importance of the conjunctival membranes and the exchange of oxygen between the environment and the cornea. Relate this to individuals who wear extended-wear contact lenses and the resulting potential for corneal hypoxia.

V. AUDIOVISUAL MATERIALS

A. Videocassettes

1. Eyes and Ears (26 min.; FHS).
2. Vision and Movement (60 min.; 1984; KSU).
3. To Hear A Pin Drop (30 min.; HR).
4. Cataracts (10 min.; C; Sd; PLP).
5. Eye Dissection and Anatomy (16 min.; C; Sd; 1990; FHS).
6. Hearing (19 min.; C; Sd; 1987; FHS).
7. The Senses (29 min.; C; SD; 1978; IM).
8. The Sound of Silence (26 min.; C; Sd; 1990; FHS).
9. What Smells? (60 min.; C; Sd; 1992; FHS).
10. What the Nose Knows (26 min.; C; Sd; 1989; FHS).

B. Film: 16 mm

1. Human Body: Sense Organs (19 min.; 1965; COR/KSU).
2. I Am Joe's Eye (26 min.; 1984; PYR/KSU).

3. Colors and How We See Them (22 min.; 1977; PYR).
4. Eye Emergency (23 min.; 1978; PYR/KSU).
5. An Introduction to Visual Illusions (16 min.; 1968; PYR/KSU).
6. Visual Illusions (25 min.; 1980; MG/KSU).
7. Across the Silence Barrier (57 min.; 1977; TLF/KSU).
8. Kevin (15 min.; 1969; CHUR/KSU).
9. My Friends Call Me Tony (12 min.; 1975; McG/KSU).

C. TRANSPARENCIES: 35 MM (2x2)

1. PAP Slide Set (Slides 68-75).
2. AHA Slide Set.
3. Histology of Sensory Organs (EIL).
4. Eyes and Their Functions (20 Slides; EIL).
5. Ears and Their Function (20 Slides; EIL).
6. Touch, Taste and Smell (20 Slides; EIL).
7. Organs of Special Senses (40 Slides; CARO).

D. OVERHEAD TRANSPARENCIES

1. PAP Transparency Set (Trs. 16.1a&b, 16.2 a-c, 16.4a&b, 16.5, 16.7, 16.9a-c, 16.10a-g, 16.11-16.13, 16.15-16.17, 16.18, 16.19a-d, 16.20, 16.21a&b).
2. The Eye (CARO).
3. The Ear (CARO).
4. The Hearing Process (CARO).
5. Structure of the Eye (TSED).
6. Eye (HSC).
7. Taste and Smell (CARO).
8. Eye (HSC).
9. Eye and Ear (GAF).
10. Taste and Smell (CARO).

E. COMPUTER SOFTWARE

1. Eye Probe (Apple; SC-175033; PLP).
2. Ear Probe (Apple; SC-175031; PLP).
3. Senses: Physiology of Human Perception (Apple; IBM; MAC; PLP).
4. Body Language: The Nervous System (Apple II Series; ESP).
5. Dynamics of the Human Ear (Apple; IBM; EI).
6. Dynamics of the Human Eye (Apple; IBM; EI).
7. The Eye (Apple IIs; Queue).

CHAPTER 13

ENDOCRINE SYSTEM

CHAPTER AT A GLANCE

- ENDOCRINE GLANDS
- COMPARISON OF NERVOUS AND ENDOCRINE SYSTEMS
- OVERVIEW OF HORMONAL EFFECTS
- MECHANISM OF HORMONE ACTION
    - *Overview*
    - *Receptors*
    - *Lipid-soluble Hormones*
    - *Water-soluble Hormones*
- CONTROL OF HORMONAL SECRETIONS: FEEDBACK CONTROL
    - *Levels of Chemical in the Blood*
    - *Nerve Impulses*
    - *Chemical Secretions from the Hypothalamus*
- PITUITARY GLAND
    - *Anterior Pituitary*
        - *Human Growth Hormone*
        - *Thyroid-Stimulating Hormone*
        - *Adrenocorticotropic Hormone*
        - *Follicle-Stimulating Hormone*
        - *Leutinizing-Hormone*
        - *Prolactin*
        - *Melanocyte-Stimulating Hormone*
    - *Posterior Pituitary*
        - *Oxytocin*
        - *Antidiuretic Hormone*
- THYROID GLAND
    - *Function and Control of Thyroid Hormones*
    - *Calcitonin*
- PARATHYROID GLANDS
    - *Parathyroid Hormone*
- ADRENAL GLANDS
    - *Adrenal Cortex*
        - *Mineralcorticoids*
        - *Glucocorticoids*
        - *Androgens*
    - *Adrenal Medulla*
        - *Epinephrine and Norepinephrine*
- PANCREAS
    - *Glucagon*
    - *Insulin*

- Ovaries and Testes
- Pineal Gland
- Thymus Gland
- Other Endocrine Tissues
- Stress and General Adaptation Syndrome
- *Stressors*
- *Alarm Reaction*
- *Resistance Reaction*
- *Exhaustion*
- *Stress and Disease*
- Common Disorders
- Wellness Focus: Diabetes and Wellness Life Style

I. CHAPTER SYNOPSIS

Students are introduced to a comparison between nervous and endocrine systems and their homeostatic regulation. Hormones, the basic secretions of the endocrine system, are then defined and categorized into five broad areas. This is followed by the chemical classification of hormones into amines, proteins and peptides, and steroids. The mechanism of hormonal action is then considered with a discussion on receptors, plasma membrane receptor interaction, and intracellular receptor interaction. The feedback control mechanisms are discussed. The structure and function of the pituitary, thyroid, parathyroids, adrenal cortex and medulla, pancreas, ovaries, testes and thymus are outlined. Some specialized endocrine tissues, such as the placenta, are considered. The chapter concludes with a discussion of stress and the general adaptation syndrome. Among the topics considered are stressors, alarm and resistance reactions, and exhaustion. A list of common disorders associated with the endocrine system and a list of medical terms and conditions are presented.

II. LEARNING GOALS/STUDENT OBJECTIVES

1. Compare the functions of the nervous and endocrine systems in maintaining homeostasis.
2. Distinguish hormones into three general classes based on their chemistry.
3. Explain how hormones act on body cells.
4. Explain how levels of hormones in the blood are regulated.
5. Describe the location, histology, and functions of the pituitary gland.
6. Describe the location, histology, and functions of the thyroid gland.
7. Describe the location, histology, and functions of the parathyroid gland.
8. Describe the location, histology, and functions of the adrenal gland.
9. Describe the location, histology, and functions of the pancreas.
10. Describe the location, histology, and functions of the ovaries and testes.
11. Describe the location, histology, and functions of the pineal gland.
12. Describe how the body responds to stress and how stress and disease are related.

III. SAMPLE LECTURE OUTLINE

A. COMPARISON OF NERVOUS AND ENDOCRINE SYSTEMS

1. The nervous and endocrine systems work together to coordinate all activities of the body. The NERVOUS SYSTEM controls homeostasis through electrical impulses delivered over neurons. The ENDOCRINE SYSTEM releases chemical messengers in the form of hormones into the bloodstream.
2. The NERVOUS SYSTEM sends impulses to specific sets of cells. In contrast, the ENDOCRINE SYSTEM as a whole, sends messages to any point in the body through the bloodstream. The NERVOUS SYSTEM evokes muscular contraction and causes glands to secrete, whereas the ENDOCRINE SYSTEM brings about changes in the metabolic activities of the body.
3. NEURONS work within milliseconds. HORMONES may take up to several hours to bring about a response.
4. HORMONES have five broad actions. They regulate the chemical composition and volume of the internal environment, regulate metabolism and energy productions, adjust to emergency environmental demands, assume a role in coordination, growth, and development, and contribute to the basic processes of reproduction.
5. ENDOCRINE GLANDS comprise the endocrine system. In contrast to exocrine glands, which secrete their substance onto the surface of the body, endocrine glands secrete hormones directly into the blood stream.
6. The ENDOCRINE HORMONES are grouped into three classes: AMINES, PROTEINS and PEPTIDES, and STEROIDS.

B. OVERVIEW OF HORMONAL EFFECTS

1. The actions and effects of hormones can be categorized into seven broad areas. These are:
   - HORMONES REGULATE THE CHEMICAL COMPOSITION AND VOLUME OF THE INTERNAL ENVIRONMENT.
   - HORMONES HELP REGULATE METABOLISM AND ENERGY BALANCE.
   - HORMONES HELP REGULATE THE CONTRACTION OF SMOOTH AND CARDIAC MUSCLE FIBERS AND SECRETION BY GLANDS.
   - HORMONES HELP MAINTAIN HOMEOSTASIS DESPITE EMERGENCY ENVIRONMENTAL DISRUPTIONS SUCH AS INFECTION, TRAUMA, EMOTIONAL STRESS, AND DEHYDRATION.
   - HORMONES REGULATE CERTAIN ACTIVITIES OF THE IMMUNE SYSTEM.
   - HORMONES PLAYA ROLE IN THE SMOOTH, SEQUENTIAL INTEGRATION OF GROWTH AND DEVELOPMENT.
   - HORMONES CONTRIBUTE TO THE BASIC PROCESSES OF REPRODUCTION, INCLUDING GAMETE PRODUCTION, FERTILIZATION, NOURISHMENT OF THE EMBRYO AND FETUS, AND DELIVERY.

C. CHEMISTRY OF HORMONES

1. Hormones are categorized into three chemical classes. These are LIPID DERIVATIVES, AMINO ACID DERIVATIVES, and PEPTIDES and PROTEINS.
2. There are two types of hormones that are derived from lipid derivatives. These are STEROID HORMONES, derived from cholesterol, and EICOSANOIDS, derived from fatty acids.
3. The two families of eicosanoids are PROSTAGLANDINS and LEUKOTRINES, which act as local hormones in the body.

4. The lipid derivatives exhibit such functions as regulating the normal physiology of smooth muscle, blood flow, reproduction, platelet function, nerve impulse transmission, immune responses, promoting fever, and tissue inflammation.
5. The AMINO ACID DERIVATIVES are the simplest of the hormones and are derived from amino acids.
6. The PEPTIDES and PROTEINS consist of chains of amino acids, anywhere from 3 to 200 in length.
7. All of these hormones function in maintaining homeostasis by changing the physiological activity of cells.

D. MECHANISM OF HORMONE ACTION

1. HORMONES are released on demand. The amount of hormone released by a gland is determined by the body's need for the hormone at any given time.
2. The hormone's principal action site is called the TARGET AREA, which is comprised of TARGET CELLS. These cells contain specific RECEPTORS that will bind with one or more hormones. Hormones will only effect those cells with specific target receptors.
3. Although a given hormone will only bind with specific cells, different types of cells can possess receptors for the same hormone, and their responses may be different from one another.
4. When a hormone binds to a cell's receptor, a chain of events is initiated that will cause target cells to alter their rate of function. The type of alteration produced depends upon whether the hormone binds with plasma membrane receptors or with intracellular receptors.
5. PLASMA MEMBRANE RECEPTORS are affected by water-soluble hormones that are amines or proteins and peptides, because they cannot penetrate the lipid layer of the plasma membrane.
6. When water-soluble hormones are released into the blood, they convey a specific message to the target cell. The hormone is referred to as the FIRST MESSENGER.
7. Since they are not lipid-soluble, they rely on a SECOND MESSENGER to convey the hormonal message to the inside of the cell in order to get a hormonal response.
8. CYCLIC ADENOSINE MONOPHOSPHATE (cAMP) is the best known SECOND MESSENGER and its synthesis increases when a water-soluble hormone attaches to a plasma membrane receptor.
9. This interaction causes the internal portion of the plasma membrane to release ADENYLATE CYCLASE, which will convert ATP into cAMP.
10. cAMP does not directly produce the given physiological response but activates a series of enzymes, called protein KINASES, which will provoke the cellular response.
11. CELLULAR RESPONSES include the regulation of enzyme synthesis, the induction of secretion, activation of protein synthesis, and the alteration of plasma membrane permeability.
12. cAMP is rapidly degraded by an enzyme known as PHOSPHODIESTERASE (PDE). CALCIUM ions and CYCLIC GUANOSINE MONOPHOSPHATE (cGMP) may also act as second messengers.
13. Lipid-soluble hormones, such as the steroids and thyroid hormones, alter cell function by activating genes. They pass through the membrane and interact with intracellular receptors in the nucleus. Activated genes will code for certain proteins indicated by the hormone.

E. LIPID-SOLUBLE HORMONES

1. These hormones are able to diffuse through the phospholipid layer of the membrane to gain access into the cell. Their mechanism of action is as follows:
   - A LIPID-SOLUBLE HORMONE DIFFUSES FROM THE BLOOD, THROUGH THE INTERSTITIAL FLUID, AND THROUGH THE PHOSPHOLIPID BILAYER OF THE PLASMA MEMBRANE INTO THE CELL.

- If the cell is a target cell, then the hormone binds and activates receptors located within the cell. The activated receptor will turn on or turn off the genes.
- Newly formed RNA leaves the nucleus, and enters the cytosol, and directs the synthesis of new proteins.
- The new proteins alter the cell's activity and cause the physiological responses of that cell.

F. Water-Soluble Hormones

1. The receptors for water-soluble hormones are the plasma membrane integral proteins.
2. Since the hormone delivers the message to the cell membrane, it is called the first messenger.
3. A second messenger is needed to relay the message inside the cell where hormone stimulated responses can take place.
4. One second messenger is called cyclic AMP and is synthesized from ATP. Cyclic AMP can alter the cell's function in several ways. The mechanism of action is as follows:
   - A water-soluble hormone diffuses from the blood and binds to its receptor in a target cell's plasma membrane. This binding converts ATP to cyclic AMP.
   - Cyclic AMP (the second messenger) causes the activation of several enzymes.
   - Enzyme catalyze reactions produce physiological responses.
   - After a period of time, the cyclic AMP is deactivated and the cell's response in turned off.

G. Control of Hormone Secretions: Feedback Control

1. Most hormones are released on demand in short bursts. A negative feedback control mechanism prevents the overproduction or underproduction of a hormone.
2. Hormone secretions are controlled by levels of the circulating hormone itself, nerve impulses, and releasing factors of regulating hormones.
3. Certain hormones are regulated according to the serum levels of blood chemicals such as insulin, glucose, and aldosterone.
4. Hormones may also be regulated through chemical secretions of the hypothalamus via releasing factors. If the structure of the secretion is known, it is called a regulating hormone; if it is not known, it is called a regulating factor.

H. Anterior Pituitary

1. The pituitary gland sits in the sella turcica of the sphenoid bone. It is divided into the anterior pituitary (adenohypophysis) and posterior pituitary (neurohypophysis). The pars intermedia, an avascular zone, exists between the two areas.
2. The anterior pituitary releases hormones that regulate a wide range of bodily activities and are activated or inhibited by regulating hormones released by the hypothalamus.
3. Seven hormones can be released by the anterior pituitary with proper stimulation from the hypothalamus. These are human growth hormone (hGH), prolactin (PRL), adrenocorticotropic hormone (ACTH), melanocyte-stimulating hormone (MSH), thyroid-stimulating hormone (TSH), follicle-stimulating hormone (FSH), and leutinizing hormone.
4. Human growth hormone (hGH) stimulates body growth by acting on skeletal muscles and bone tissue. It acts indirectly by causing the liver to secrete small proteins called somatomedians which mediate its effects. Two regulating hormones control hGH. These are

GROWTH HORMONE RELEASING FACTOR (GHRH) and GROWTH HORMONE INHIBITING FACTOR (GHIH). Both are released by the hypothalamus.

5. THYROID-STIMULATING HORMONE (TSH) stimulates the production and secretion of hormones from the thyroid gland and its release is controlled by THYROTROPIN RELEASING HORMONE (TRH) from the hypothalamus.
6. ADRENOCORTICOTROPIC HORMONE (ACTH) controls the production and secretion of certain adrenal hormones. The release of CORTICOTROPIN RELEASING HORMONE (CRH) by the hypothalamus regulates the release of ACTH.
7. FOLLICLE-STIMULATING HORMONE (FSH) functions to control the development of the male and female gametic cells in the testes and ovaries respectively. It is under the control of the GONADOTROPIN RELEASING HORMONE (GnRH) released by the hypothalamus.
8. LEUTINIZING HORMONE (LH) together with FSH stimulates ovulation and the formation of the corpus luteum in the female ovary. The CORPUS LUTEUM secretes and releases PROGESTERONE which, together with ESTROGEN, prepares the uterus for implantation. The release of LH is also under the control of GnRH.
9. PROLACTIN (PRL) together with other female hormones initiates milk secretion from the mammaries. PRL has both an inhibitory and excitatory control system. The inhibitory factor released by the hypothalamus is PROLACTIN INHIBITORY FACTOR (PIF), and the excitatory factor is PROLACTIN RELEASING FACTOR (PRF).
10. MELANOCYTE-STIMULATING HORMONE (MSH) has an unknown role in humans, but administration of this hormone will cause darkening of the skin. Secretion of MSH is stimulated by a hypothalamic regulating factor called MELANOCYTE-STIMULATING HORMONE RELEASING FACTOR (MRF). It is inhibited by a MELANOCYTE-STIMULATING HORMONE INHIBITING FACTOR (MIF).

I. POSTERIOR PITUITARY

1. The POSTERIOR PITUITARY stores two hormones, OXYTOCIN (OT) and ANTIDIURETIC HORMONE (ADH), which are made in the hypothalamus.
2. OXYTOCIN stimulates the contraction of the uterus and the ejection of milk. It is controlled by uterine distention and lactation.
3. ANTIDIURETIC HORMONE stimulates water reabsorption by the kidneys and arteriole vasoconstriction. It is regulated by water concentration in the bloodstream and detected by receptors located in the hypothalamus.

J. THYROID GLAND

1. The THYROID GLAND is located inferior to the larynx and contains two lobes.
2. The thyroid consists of follicles composed of FOLLICULAR CELLS, which secrete thyroid hormones THYROXIN (T-4) and TRIIODOTHYRONINE (T-3), and PARAFOLLICULAR CELLS, which secrete CALCITONIN (CT).
3. THYROID HORMONES regulate the metabolism of organic molecules, energy balance, growth and development, and activation of the nervous system. Secretion is mediated through the action of both the hypothalamus and anterior pituitary.
4. CALCITONIN (CT) functions to lower serum levels of calcium and phosphate by depositing mineral salts into bone tissue. Secretion is controlled by the level of calcium in the blood.

K. PARATHYROIDS

1. The PARATHYROIDS are located on the posterior aspect of the thyroid gland. There is usually a pair on either side.
2. The parathyroids consist of PRINCIPAL (CHIEF) CELLS and OXYPHIL CELLS each which secrete PARATHYROID HORMONE (PTH).
3. PARATHYROID HORMONE regulates the homeostasis of calcium and phosphate by increasing blood calcium levels and decreasing blood phosphate levels. The secretion is controlled by calcium levels in the blood.

L. ADRENAL GLANDS

1. The ADRENAL GLANDS are positioned superior to the kidneys and consist of an outer MEDULLA and inner CORTEX.
2. The CORTEX is subdivided into three zones. The OUTER ZONE secretes hormones classified as MINERALCORTICOIDS, the MIDDLE ZONE secretes hormones classified as GLUCOCORTICOIDS, and the INNER ZONE secretes hormones classified as GONADOCORTICOIDS.
3. The MINERALCORTICOIDS, such as ALDOSTERONE, increase sodium and water reabsorption, decrease potassium reabsorption, and regulate sodium and potassium levels in the blood.
4. MINERALCORTICOID SECRETION is controlled by the RENIN-ANGIOTENSIN PATHWAY, which is activated by the release of the enzyme renin by kidney tubular cells called the JUXTAGLOMERULAR CELLS. The release of renin is controlled by decreases in mean arterial blood pressure, a decrease in extracellular fluid volume, or a decrease in sodium ion concentration in the extracellular fluid.
5. The GLUCOCORTICOIDS, such as CORTISONE and CORTISOL, promote the metabolism of organic materials, provide resistance to stress, and serve as anti-inflammatory agents. Secretion is controlled by CORTICOTROPIN RELEASING HORMONE (CRF)
6. The GONADOCORTICOIDS such as ESTROGEN and ANDROGEN, have minimal effects on the human.
7. The ADRENAL MEDULLA contains CHROMAFFIN CELLS, which synthesize and secrete EPINEPHRINE and NOREPINEPHRINE. These are released under stress by direct innervation from the ANS.

M. PANCREAS

1. The PANCREAS is both an exocrine and endocrine organ. In its exocrine capacity, it functions to produce digestive enzymes. In its endocrine capacity, it secretes GLUCAGON, INSULIN, and GROWTH HORMONE INHIBITING HORMONE (GHIH).
2. The PANCREAS is a flattened organ posterior and inferior to the stomach.
3. Histologically, it is comprised of patches of cells called the ISLETS OF LANGERHANS. Three cell types can be identified: ALPHA, BETA, and DELTA.
4. The ALPHA CELLS secrete GLUCAGON, which functions to increase the blood level of glucose. Its secretion is controlled by glucose levels in the blood.
5. The BETA CELLS secrete INSULIN, which functions to decrease blood levels of glucose and increase tissue utilization of glucose. Secretion of insulin is controlled by levels of glucose in the blood.
6. The DELTA CELLS secrete GHIH, which functions to inhibit the secretions of insulin and glucagon.

N. OVARIES AND TESTES

1. The OVARIES are glands located in the pelvic cavity of the female and produce sex hormones related to the development and maintenance of female sexual characteristics, menstrual cycle, pregnancy, lactation, and normal reproductive functions. The chief hormones released by the ovaries are ESTROGEN and PROGESTERONE.
2. The TESTES are glands located external to the body of the male in a sac called the scrotum. They produce sex hormones related to the development and maintenance of male sexual characteristics and normal reproductive functions. The chief hormone is TESTOSTERONE.

O. THE PINEAL GLAND

1. The PINEAL GLAND is attached to the roof of the third ventricle of the brain near the hypothalamus. It accumulates calcium about the time of puberty. These deposits are known as BRAIN SAND.
2. The PINEAL GLAND secretes MELATONIN, which appears to inhibit reproductive activities by inhibiting gonadotropic hormones.

P. THYMUS

1. The THYMUS GLAND produces several hormones related to immunity. THYMOSIN, THYMIC HUMORAL FACTOR (THF), THYMIC FACTOR (TF), and THYMOPOIETIN promote the development and maturation of T-cells, a type of white blood cell involved in immunity.

Q. OTHER ENDOCRINE TISSUES

1. Body tissues, other than those classified as endocrine glands, also contain endocrine tissue and secrete hormones.
2. The GASTROINTESTINAL TRACT synthesizes and secretes GASTRIN, SECRETIN, CHOLECYSTIKININ, ENTEROCRININ, VILLIKININ, and GASTRIC INHIBITORY PEPTIDES.
3. The PLACENTA produces HUMAN CHORIONIC GONADOTROPIN, ESTROGEN, PROGESTERONE, and RELAXIN, all related to pregnancy.
4. The KIDNEYS release RENAL ERYTHROPOIETIC FACTOR that causes the production of red blood cells.
5. The ATRIA of the heart produce ATRIAL NATRIURETIC FACTOR, which helps to lower blood pressure.

R. EICOSAMOIDS

1. PROSTAGLANDINS and LEUKOTRINES act as local hormones in most body tissues by altering the production of second messengers.
2. Prostaglandins have a wide range of biological activity in normal physiology and pathology.

S. STRESS AND THE GENERAL ADAPTATION SYNDROME

1. If a STRESS is extreme or unusual, it triggers a range of physiological responses called the GENERAL ADAPTATION SYNDROME (GAS). This response does not maintain homeostasis.

2. The stimuli that produce the general adaptation syndrome are referred to as STRESSORS and can include surgical operations, poisons, infections, fever, and strong emotional responses.
3. There are three pathways involved in the general adaptation reaction: the ALARM REACTION, RESISTANCE REACTION, and EXHAUSTION.
4. The ALARM REACTION is the body's initial response to a stressor and is initiated by the hypothalamus and adrenal medulla. The responses are immediate and short-lived flight-or-fight responses that increase circulation and energy production, and decrease non-essential activities.
5. The RESISTANCE REACTION is the second phase in the response to a stressor and is initiated by the hypothalamus and regulating hormones.
6. The REGULATING HORMONES are CRH, GHRH, and TRH. These create long-term reactions and accelerated catabolism to provide energy. Glucocorticoids are produced in large quantities during this phase.
7. EXHAUSTION is the end result of dramatic changes during the alarm and resistance reactions. Exhaustion is generally caused by the depletion of potassium, adrenal glucocorticoids, and weakened organs.

Renin-Angiotensin Pathway

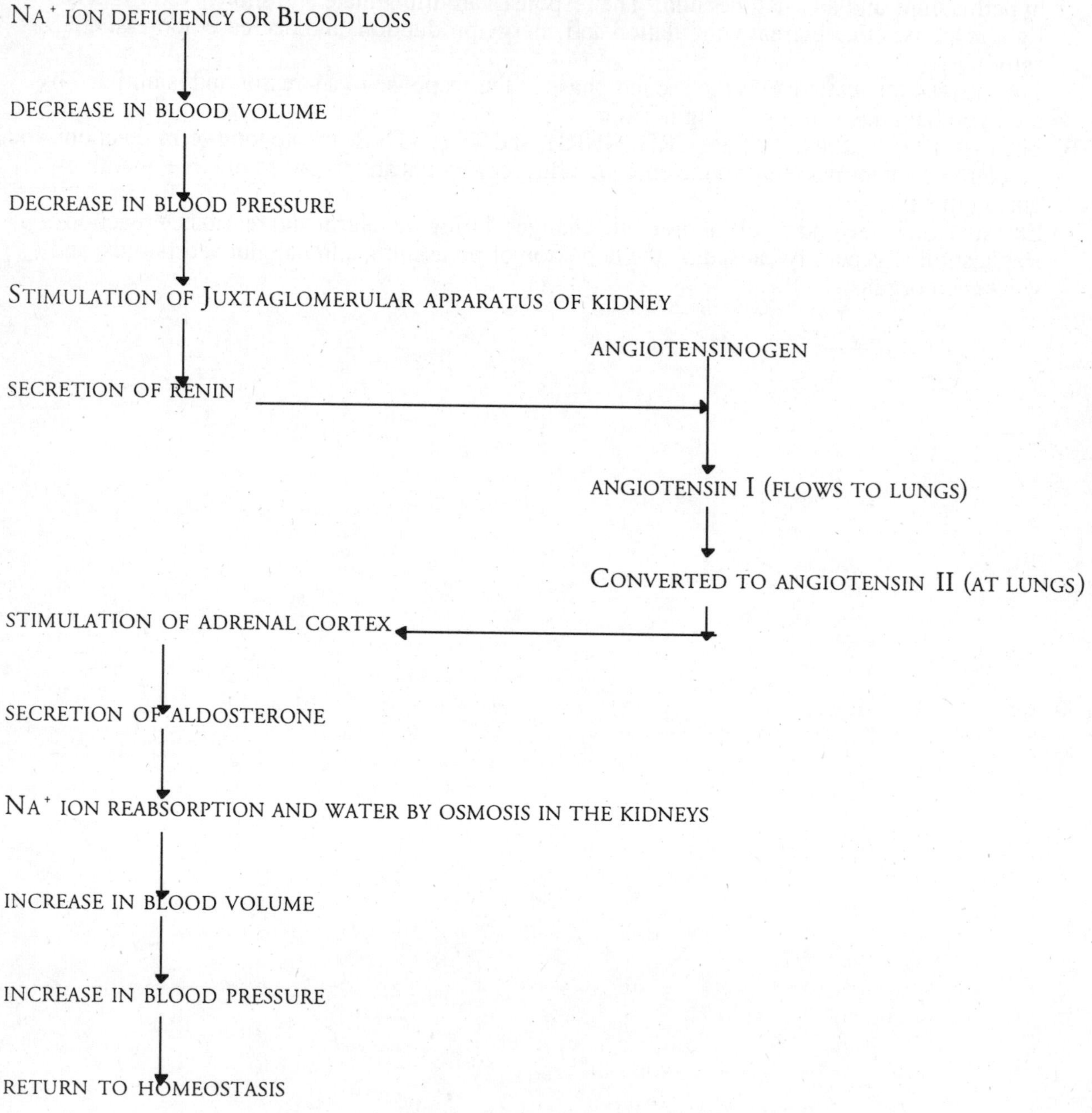

Na^+ ion deficiency or Blood loss

↓

decrease in blood volume

↓

decrease in blood pressure

↓

Stimulation of Juxtaglomerular apparatus of kidney

↓

secretion of renin → angiotensinogen

↓

angiotensin I (flows to lungs)

↓

Converted to angiotensin II (at lungs)

↓

stimulation of adrenal cortex

↓

secretion of aldosterone

↓

Na^+ ion reabsorption and water by osmosis in the kidneys

↓

increase in blood volume

↓

increase in blood pressure

↓

return to homeostasis

IV. TEACHING TIPS AND SUGGESTIONS

A. Helpful Hints

1. Usage of transparencies to locate the numerous endocrine glands is suggested.
2. It is useful to break up the material into workable units by considering one endocrine organ at a time and learning the hormones and the acronyms. Outline clearly the level of detail you expect your students to know. Consider a few quizzes on a few organs at a time rather than a single comprehensive endocrine examination.
3. To illustrate homeostatic imbalances, it may be useful to discuss the common abnormalities associated with hyposecretions and hypersecretions of some selected glands.

B. Essays

1. An individual with a tumor of the hypothalamus is passing a large volume of urine. What relationship exists between the hypothalamus and urine production? What hormones are probably being undersecreted?
2. List the numerous glands and their principle hormones and acronyms. List the relationship between the hypothalamus, the anterior pituitary, and the chosen gland.

V. AUDIOVISUAL MATERIALS

A. Videocassettes

1. Messengers (26 min.; FHS).
2. The Endocrine System (48 min.; CR).
3. Regulatory Systems (30 min.; C; Sd; PLP).
4. The Endocrine Glands (29 min.; C; Sd; 1978; IM).
5. The Endocrine System (48 min.; 1987; GA).
6. Hormones and the Endocrine System (45 min.; C; Sd; 1981; IM).

B. Films: 16 mm

1. The Human Body: Endocrine System (15 min.; 1980; COR/KSU).
2. Endocrine Glands (14 min.; 1973; KSU).
3. Prostaglandins: Tomorrow's Physiology (22 min.; C; Sd; 1974; UPFL).

C. Transparencies: 35 mm (2x2)

1. PAP Slide Set (Slides 76-79).
2. AHA Slide Set.
3. Endocrine System (Slides 157-172; McG).
4. Visual Approach to Histology: Endocrine Glands (10 Slides; FAD).
5. Hormones and the Endocrine System, Parts 1-4 (224 Slides; IBIS).
6. Stress and Disease (148 Slides; IBIS).
7. Hormones (40 Slides; CARO).
8. The Endocrine System Set (20 Slides; CARO).

D. OVERHEAD TRANSPARENCIES

1. PAP Transparency Set (Trs. 18.1-18.12, 18.13a, 18.14, 18.15, 18.17, 18.18a, 18.19a, 18.20, 18.21, 18.23a, 18.24, 18.26).
2. Endocrine System (GAF).
3. Endocrine System- Unit 10 (11 Transparencies; RJB).
4. Activity of a Tropic Hormone (K&E).

E. COMPUTER SOFTWARE

1. Hormones (Apple II Series; C4556; EIL).
2. Blood Sugar (Apple; SC-390784; PLP).
3. The Endocrine System (Apple II Series; ESP).
4. Biochemistry of Hormones (Apple II Series; CARO).
5. Dynamics of the Human Endocrine System (Apple; IBM; EIL).
6. Nervous and Hormonal Systems (Apple IIs; IBM; MAC; Queue).

CHAPTER 14

THE CARDIOVASCULAR SYSTEM: BLOOD

CHAPTER AT A GLANCE

- FUNCTIONS OF BLOOD
- PHYSICAL CHARACTERISTIC OF BLOOD
- COMPONENTS OF BLOOD
- *Formed Elements*
- *Formation of Blood Cells*
- *Erythrocytes (Red Blood Cells)*
- *Leukocytes (White Blood Cells)*
- *Platelets*
- *Plasma*
- HEMOSTASIS
- *Vascular Spasm*
- *Platelet Plug Formation*
- *Coagulation*
- *Extrinsic Pathway*
- *Intrinsic Pathway*
- *Clot Retraction and Fibrinolysis*
- *Hemostatic Control Mechanisms*
- *Clotting in Blood Vessels*
- GROUPING (TYPING) OF BLOOD
- *ABO*
- *Rh*
- COMMON DISORDERS
- MEDICAL TERMINOLOGY AND CONDITIONS
- WELLNESS FOCUS: SAFE TRANSFUSIONS- FACT OR DELUSION?

I. CHAPTER SYNOPSIS

The major concern of this chapter is to analyze the origin, structure, and function of blood and its relationship in the maintenance of homeostasis. This is developed through a study of the origin and functions of the formed elements in the blood, and a comparison of the location and composition of plasma. Consideration is also given to hemostasis and blood clotting as well as hemostatic control mechanisms. The ABO and Rh blood grouping systems are considered. The chapter concludes with a list of common medical disorders and medical terms and conditions associated with the blood.

II. LEARNING OBJECTIVES/STUDENT GOALS

1. List the components and functions of blood components.
2. List the components and functions of blood plasma.

III. SAMPLE LECTURE OUTLINE

A. FUNCTIONS OF BLOOD

1. Blood functions to TRANSPORT oxygen, carbon dioxide, nutrients, water, hormones, and heat. It also functions to regulate pH, body temperature, and water content of cells.
2. HEMOSTASIS and BLOOD CLOTTING prevent blood loss. Phagocytic white blood cells and specialized proteins fight against microbes and toxins.

B. PHYSICAL CHARACTERISTICS OF BLOOD

1. Blood is a specialized CONNECTIVE TISSUE of the cardiovascular system.
2. The VISCOSITY of blood is 4.5 to 5.5 (water has a viscosity of 1.0). Blood temperature is 38° Celsius. Blood has a pH of 7.35 to 7.45 and a salt (NaCl) concentration of 0.90%. Blood constitutes about 8% of the body's weight.

C. COMPONENTS OF BLOOD

1. About 45% of the blood volume consists of FORMED ELEMENTS including ERYTHROCYTES (red blood cells), LEUKOCYTES (white blood cells), and THROMBOCYTES (platelets). The remaining 55% of the blood volume in the liquid portion of the blood is called PLASMA.
2. Blood cells are formed by a process called HEMOPOIESIS. This occurs in the RED BONE MARROW, or MYELOID TISSUE, which forms the red blood cells, granular leukocytes, and platelets. The LYMPHOID TISSUE and myeloid tissue produce agranular leukocytes.
3. All blood cells originate from immature HEMOCYTOBLASTS that undergo differentiation to become mature blood cells.

D. FORMED ELEMENTS

1. The formed elements of blood are:
   - ERYTHROCYTES OR RED BLOOD CELLS
   - LEUKOCYTES OR WHITE BLOOD CELLS
   - GRANULAR LEUKOCYTES OR THE GRANULOCYTES
     - ∞ NEUTROPHILS
     - ∞ EOSINOPHILS
     - ∞ BASOPHILS
   - AGRANULAR LEUKOCYTES OR AGRANULOCYTES
     - ∞ LYMPHOCYTES
     - ∞ MONOCYTES
   - PLATELETS OR THROMBOCYTES

E. ERYTHROCYTES

1. ERYTHROCYTES, also known as red blood cells or red blood corpuscles, are anucleate, biconcave discs containing the pigment HEMOGLOBIN.
2. The function of red blood cells is to TRANSPORT OXYGEN and CARBON DIOXIDE. Hemoglobin molecules are specialized components of the red blood cell plasma membrane which combine with oxygen and carbon dioxide in this transport process.
3. Red blood cells live approximately 120 days. Worn out cells are broken down by phagocytosis in the spleen, liver, and bone marrow.
4. The GLOBIN component is broken down into amino acids and may be reused by other cells for protein synthesis.
5. The HEME is broken down into iron and biliverdin. The iron is stored in the liver as ferritin and hemosiderin. Biliverdin is converted to bilirubin and then transported to the liver to be excreted in the bile.
6. Red blood cells are produced by a process called ERYTHROPOIESIS in the red bone marrow. The average individual contains approximately 5 million RBC per cubic millimeter.
7. Erythropoiesis is triggered by a reduced supply of oxygen to the cells, a condition known as hypoxia. Specialized kidney and liver cells detect decreases in oxygen and release erythropoietic factor and globulin, respectively. The two components form a hormone called ERYTHROPOIETIN. Erythropoietin stimulates the myeloid tissue in bone marrow to increase the production of red blood cells.
8. The rate of erythropoiesis is measured by a procedure called the RETICULOCYTE COUNT. The HEMATOCRIT measures the percentage of red blood cells in whole blood and is represented as a percentage with 47% and 42% being normal for males and females, respectively.

F. LEUKOCYTES

1. LEUKOCYTES, or white blood cells, are nucleated cells and are classified into two major divisions. These are GRANULAR and AGRANULAR LEUKOCYTES.
2. The GRANULAR LEUKOCYTES include NEUTROPHILS, BASOPHILS, and EOSINOPHILS, collectively known as the POLYMORPHONUCLEAR LEUKOCYTES. The AGRANULAR cells include the LYMPHOCYTES and MONOCYTES.
3. Leukocytes have surface proteins called HUMAN LEUKOCYTE-ASSOCIATED ANTIGENS (HLA). They are also known as HISTOCOMPATIBILITY ANTIGENS and are unique in each individual, except for identical twins.
4. The general function of leukocytes is to combat infection and inflammation. The neutrophils and monocytes accomplish this through phagocytosis. They are often referred to as wandering macrophages.
5. EOSINOPHILS combat the effects of histamine in allergic reactions, phagocytize antigen-antibody complexes, and combat parasitic worms.
6. BASOPHILS release HEPARIN, HISTAMINE, and SERATONIN in allergic reactions that intensify the inflammatory response.
7. Lymphocytes respond to foreign substances called ANTIGENS and differentiate into PLASMA CELLS that produce ANTIBODIES. Antibodies attach to and inactivate the antigens. COMPLEMENT, a series of enzymatic reactions in the plasma, then acts to destroy the antigen.
8. White blood cells usually live from a few hours to several days. The normal white blood count is 5,000 to 10,000 per cubic millimeter of blood.

9. White blood cells are developed under the influence of substances called COLONY-STIMULATING FACTORS (CSF) found in red bone marrow and lymphoid tissue.

G. THROMBOCYTES

1. THROMBOCYTES, or platelets, are disc-shaped structures without nuclei.
2. They are formed from MEGAKARYOCYTES in the red bone marrow and are involved in blood clotting.
3. PLATELETS have a normal life span which averages only five to nine days and normal blood contains 250,000 to 400,000 platelets per cubic millimeter.

H. PLASMA

1. The liquid portion of the blood is called PLASMA and consists of 91.5% water and 8.5% solutes.
2. The principal solutes include plasma proteins (albumins, globulins, and fibrinogen), non-protein nitrogenous substances, nutrients, hormones, respiratory gases, and electrolytes.

I. HEMOSTASIS OF THE BLOOD

1. HEMOSTASIS refers to the prevention of blood loss through the reduction of blood flow. It involves VASCULAR SPASM, PLATELET PLUG FORMATION, and BLOOD COAGULATION or CLOTTING.
2. In a VASCULAR SPASM, the smooth muscle of a damaged blood vessel wall constricts to prevent blood loss.
3. PLATELET PLUG FORMATION involves the clumping of platelets to impede the loss of blood.
4. A BLOOD CLOT is a network of insoluble protein (fibrin) in which the formed elements of the blood become trapped.
5. The chemicals involved in blood coagulation are known as coagulation factors and are classified as plasma or platelet coagulation factors, depending upon where they originate.
6. The CLOTTING MECHANISM is a cascade system which can be broken into three stages: FORMATION OF PROTHROMBIN ACTIVATOR, CONVERSION OF PROTHROMBIN TO THROMBIN, and the CONVERSION OF FIBRINOGEN TO FIBRIN.
7. The formation of PROTHROMBIN ACTIVATOR is initiated by the interplay of two mechanisms: the EXTRINSIC and INTRINSIC PATHWAYS of blood clotting.
8. The EXTRINSIC PATHWAY occurs rapidly and is so named because the formation of prothrombin activator is initiated by tissue factor (TF), also called thromboplastin, which is found on the surfaces of cells found outside of the cardiovascular system.
9. The clotting mechanism consists of three stages.
10. In STAGE 1, damaged tissues release tissue factor which after several reactions involving calcium, is converted into prothrombin activator.
11. In STAGE 2, prothrombin activator and calcium ions convert prothrombin into thrombin.
12. In STAGE 3, thrombin converts fibrinogen to fibrin.
13. Thrombin has a positive feedback effect in accelerating the formation of prothrombin activator.
14. The INTRINSIC PATHWAY is more complex and operates more slowly than the extrinsic pathway. This pathway is so named because the formation of prothrombin activator is initiated by a tissue factor found on the surfaces of endothelial cells that line blood vessels and is released into the bloodstream.

15. The INTRINSIC PATHWAY is triggered when blood comes in contact with the tissue factor from damaged endothelial cells. The tissue factor damages the platelets which release phospholipids.
16. Several reactions and the presence of calcium ions provide the generation of prothrombin activator. This completes the intrinsic pathway and stage 1 of blood clotting. Stages 2 and 3 are similar to those in the extrinsic pathway.
17. Normal coagulation involves two additional events after clot formation: clot retraction and fibrinolysis.
18. CLOT RETRACTION is the tightening of the fibrin clot, which decreases the possibility of further blood loss.
19. FIBRINOLYSIS is the dissolution of the blood clot. When a clot is formed, an inactive enzyme called plasminogen is activated to plasmin. Plasmin acts to digest the fibrin threads and inactivate fibrinogen, prothrombin, and coagulation factors.

J. GROUPING (TYPING) OF BLOOD

1. The surfaces of red blood cells contain genetically determined blood group ANTIGENS. The plasma may contain genetically determined antibodies against the blood group antigens which they do not possess. The ABO and Rh blood group systems are based on antigen-antibody responses.
2. In the ABO system, AGGLUTINOGENS (antigens) A and, or B, determine blood type. Plasma contains AGGLUTININS (antibodies), designated as a and, or b, which will clump any agglutinogen that is foreign to the individual.

BLOOD TYPE	AGGLUTINOGENS	AGGLUTININS
A	A	b
B	B	a
AB	A & B	none
O	none	a & b

3. Blood typing is essential for the safe transfusion of blood and may be used in disproving paternity, linking individuals to crimes, and as part of anthropological studies.
4. In the Rh system, individuals whose erythrocytes have Rh agglutinogens are classified as Rh+; those who lack them are classified as Rh-.
5. A disorder due to Rh incompatibility between mother and fetus is called hemolytic disease of the newborn (HDN), or erythroblastosis fetalis.

K. COMMON DISORDERS

1. ANEMIA is any condition in which the oxygen carrying capacity of the blood is impaired. Some classifications are: PERNICIOUS ANEMIA, HEMORRHAGIC ANEMIA, HEMOLYTIC ANEMIA, APLASTIC ANEMIA, and SICKLE-CELL ANEMIA.
2. POLYCYTHEMIA refers to an abnormal increase in the number of red blood cells.
3. INFECTIOUS MONONUCLEOSIS is a contagious disease affecting lymphoid tissue mainly in young adults and children and is caused by the Epstein-Barr virus.
4. LEUKEMIA is an uncontrolled production of immature leukocytes.

IV. TEACHING TIPS AND SUGGESTIONS

A. Helpful Hints

1. A slide or transparency presentation is helpful when lecturing on the white blood cells.
2. In the laboratory environment, a histological study of the numerous types of blood cells will be useful.
3. If allowed, a demonstration of hematocrit, clotting time, blood typing, hemoglobin determination, and sedimentation rate will give the student more insight into the overall characteristics of blood.

B. Essays

1. A laboratory test of an individual indicated a hematocrit of 15%. Microscopic examination of the blood reveals several distorted and ruptured red blood cells. In addition, the reticulocyte count is 2%. Based on these findings, what disorder do you think the individual may be suffering from?
2. Individuals with type AB blood can theoretically receive blood from individuals with type A, B, AB, and O blood. Conversely, individuals with type O blood can theoretically give blood to individuals with type A, B, AB, and O blood. Explain why both situations are possible on the basis of antigen-antibody responses. Can you relate this to hemolytic disease of the newborn? How could this condition be treated? How can it be prevented?

V. AUDIOVISUAL MATERIALS

A. Overhead Transparencies (PAP)

1. PAP Transparency Set (Trs. 19.1, 19.2, 19.3a&b, 19.4-19.7, 19.8a-c, 19.10 & 19.11).

1. Circulatory System: Coagulation of Blood (CARO).
2. Human Blood Cell Formation (CARO).
3. Constituents and Function of Blood (CARO).
4. Defensive Action of White Blood Corpuscles (CARO).

B. Videocassettes

1. Blood: The Vital Humor (30 mins.; 1989; H&R).
2. Blood (9 mins.; 1988; FHS).
3. Life Under Pressure (26 mins.; FHS).
4. Blood: River of Life, Mirror of Health (HRM).
5. Blood (33 mins.; PLP).
6. What Can Go Wrong? (43 mins.; 1983; PLP).
7. Blood: The Microscopic Miracle (22 mins.; 1983; EBEC/KSU).
8. Blood: The Microscopic Miracle, 2/e (22 mins.; C; Sd; 1990; GAF).
9. HB Masters Sickle Cell Anemia (23 mins.; C; Sd; 1992; FHS).
10. The Human Body-What Can Go Wrong? - Circulatory System (43 mins.; C; Sd; PLP).
11. The Life of a Red Blood Cell (10 mins.; C; Sd; 1990; KSU).

C. Films: 16 mm

1. Blood: The Microscopic Miracle (22 mins.; 1983; EBEC/KSU).
2. The Human Body: Circulatory System (16 mins.; 1980; COR/KSU).
3. How Blood Clots (13 mins.; 1969; BFA/KSU).
4. White Blood Cells (11 mins.; 1961; McG/KSU).
5. Secret of the White Blood Cell (24 mins.; NET).
6. Circulation (28 mins.; 1961; McG/KSU).
7. The Work of Blood (14 mins.; EBEC).
8. Hemoglobin (25 mins.; UIFC).

D. Transparencies: 35 mm (2x2)

1. PAP Slide Set (Slides 80-82).
2. AHA Slide Set.
3. Visual Approach to Histology: Blood and Bone Marrow (19 Slides; FAD).
4. Hematology (120 Slides; EIL).
5. Histology of Blood (EIL).
6. Blood Under the Microscope: Medical Slide Series (214 Slides; CARO).

E. Computer Software

1. Biology Program Series: Transport (Apple II; 39-8848; TRS-80 Model III; 39-8849; CARO).
2. Simulation of Hemoglobin Function (Apple II; 39-9430; CARO).
3. Dynamics and Human Circulation (Apple; IBM; EI).
4. Circulation and Respiration (Apple IIs; IBM/PC; MAC; Queue).

CHAPTER 15

THE CARDIOVASCULAR SYSTEM: HEART

CHAPTER AT A GLANCE

- LOCATION OF THE HEART
- PERICARDIUM
- HEART WALL
- CHAMBERS OF THE HEART
- GREAT VESSELS OF THE HEART
- VALVES OF THE HEART
  - *Atrioventricular (AV) Valves*
  - *Semilunar Valves*
- BLOOD SUPPLY OF THE HEART
- CONDUCTION SYSTEM OF THE HEART
- ELECTROCARDIOGRAM (ECG OR EKG)
- BLOOD FLOW THROUGH THE HEART
- CARDIAC CYCLE
  - *Phases*
  - *Timing*
  - *Sounds*
- CARDIAC OUTPUT (CO)
- HEART RATE
  - *Autonomic Control*
  - *Chemicals*
  - *Temperature*
  - *Emotions*
  - *Gender and Age*
- RISK FACTORS IN HEART DISEASE
- MEDICAL TERMINOLOGY AND CONDITIONS
- WELLNESS FOCUS: DIET AND HEART DISEASE- REDEFINING THE GOOD LIFE

I. CHAPTER SYNOPSIS

This chapter presents the student with the major structural and functional features of the heart. The principal structural features discussed are the pericardium and heart wall, chambers, vessels, and valves. The principal functional features discussed are the blood supply to the heart, the initiation and conduction of the heart beat, the electrocardiogram, blood flow through the heart, the major events of the cardiac cycle, heart sounds, cardiac output, and the regulation of heart rate. The influence of emotions, gender, and age on the heart are also considered. The chapter concludes with a discussion of the risk factors for heart disease, a list of common disorders, medical terms and conditions.

II. LEARNING GOALS/STUDENT OBJECTIVES

1. Describe the structure and function of the pericardium.
2. Describe the structure and functions of the valves of the heart.
3. Describe the clinical importance of the blood supply to the heart.
4. Explain how each heartbeat is initiated and maintained.
5. Describe the meaning and diagnostic value of an electrocardiogram.
6. Explain how blood flows through the heart.
7. Describe the phases of a cardiac cycle.
8. Identify the factors that affect heart rate.
9. List and explain the risk factors involved in heart disease.

III. SAMPLE LECTURE OUTLINE

A. LOCATION OF THE HEART

1. The heart is located between the lungs in the medial mediastinum. Approximately two-thirds of its mass is to the left of midline.
2. The APEX is the inferior pointed end of the heart. It is formed by the tip of the left ventricle and rests on the diaphragm.
3. The BASE of the heart is the superior or atrial portion. The major blood vessels are attached to the base.

B. PERICARDIUM

1. The heart is enclosed and held in position by the PERICARDIUM, which consists of an outer FIBROUS PERICARDIUM and an inner SEROUS PERICARDIUM.
2. The FIBROUS PERICARDIUM prevents over-distention of the heart, provides a tough protective membrane, and anchors the heart to the mediastinum.
3. The SEROUS PERICARDIUM is more delicate and forms a double layer around the heart. This double layer is comprised of the outer parietal pericardium and the inner visceral pericardium, also known as the EPICARDIUM.
4. Between the parietal and visceral pericardium is the PERICARDIAL CAVITY, a potential space filled with PERICARDIAL FLUID that prevents friction between the two membranes.
5. Inflammation of this membrane is known as PERICARDITIS, and is potentially fatal.

C. HEART WALL, CHAMBERS, VESSELS, AND VALVES

1. The wall of the heart has three layers: EPICARDIUM, MYOCARDIUM, and ENDOCARDIUM.
2. The EPICARDIUM is the visceral layer of the serous pericardium and is comprised of serous tissue and epithelium.
3. The MYOCARDIUM consists of cardiac muscle, which constitutes the bulk of the heart. Cardiac muscle cells are involuntary, striated and branched, and the tissue is arranged in interlacing bundles of fibers. The myocardium is responsible for the pumping action of the heart.

4. The ENDOCARDIUM is a layer of simple squamous epithelium that lines the myocardium and covers the valves of the heart and tendons that hold them. It is continuous with the inner lining of the blood vessels.

D. CHAMBERS OF THE HEART

1. The chambers of the heart include two upper ATRIA separated by an INTERATRIAL SEPTUM, and two lower VENTRICULAR chambers separated by the INTERVENTRICULAR SEPTUM.
2. The thickness of the chambers varies depending upon their function.
3. The blood flows to the heart from the SUPERIOR and INFERIOR VENA CAVAE and the CORONARY SINUS to the RIGHT ATRIUM, through the TRICUSPID VALVE to the RIGHT VENTRICLE, through the PULMONARY TRUNK to the lungs, through the PULMONARY VEINS to the LEFT ATRIUM, through the BICUSPID or MITRAL VALVE to the LEFT VENTRICLE, and out through the AORTA.

E. VALVES OF THE HEART

1. The heart contains FOUR VALVES, each composed of dense connective tissue and covered by endothelium. Each of these valves, the TWO ATRIOVENTRICULAR VALVES and the two SEMILUNAR VALVES, function to prevent the backflow of blood in the heart.
2. The two atrioventricular valves are located between the atria and ventricles. The TRICUSPID VALVE is located between right atrium and right ventricle. The BICUSPID or MITRAL VALVE is located between the left atrium and left ventricle.
3. Tendinous structures called the CHORDAE TENDINAE, and their associated PAPILLARY MUSCLES are located in the ventricular chambers. They keep the cusps of the valves pointing in the direction of blood flow to prevent eversion and backflow of blood into the atria.
4. The two SEMILUNAR VALVES prevent the backflow of blood into the heart as it leaves the ventricular chambers. The PULMONARY SEMILUNAR VALVE is located between the right ventricle and the pulmonary artery. The AORTIC SEMILUNAR VALVE is located between the left ventricle and the aorta.

F. BLOOD SUPPLY TO THE HEART

1. The flow of blood through the numerous vessels in the myocardium is called CORONARY CIRCULATION.
2. The CORONARY ARTERIES deliver oxygenated blood to the myocardium, and the coronary veins remove deoxygenated blood passing through the CORONARY SINUS, which exits into the right atrium.
3. Most heart attacks result from faulty coronary circulation. Reduced oxygen supplies to the heart muscle cells will weaken them in a condition which is known as ischemia. A clinical condition, known as angina pectoris, is a result from myocardial ischemia.

G. CONDUCTION SYSTEM

1. The conduction system consists of tissue which is specialized for the generation and distribution of action potentials that stimulate the cardiac muscle cells to contract.

2. The components of this system are the SINOATRIAL (SA) NODE, also known as the pacemaker, ATRIOVENTRICULAR (AV) NODE, ATRIOVENTRICULAR BUNDLE (BUNDLE OF HIS), BUNDLE BRANCHES, and conduction myofibers (PURKINJE FIBERS).
3. All cardiac muscle is capable of self-excitation. It spontaneously and rhythmically generates action potentials that result in the contraction of muscle. This spontaneous generation of impulses is controlled by the sinoatrial node, which generates approximately 75 action potentials per minute.
4. When an action potential is initiated by the SA NODE, it spreads over both atria, causing them to CONTRACT SIMULTANEOUSLY and depolarizing the slowly conducting AV node.
5. The AV NODE is one of the last portions of the heart to be depolarized. This allows the atria to empty their blood into the ventricles.
6. The BUNDLE OF HIS is a tract of conducting fibers that passes through the INTERVENTRICULAR SEPTUM and splits into right and left bundle branches at the apex. The BUNDLE OF HIS distributes the action potential over the ventricles, where the actual contraction of the ventricles is stimulated by the PURKINJE FIBERS, or conduction myofibers, that emerge from the bundle branches.

H. ELECTROCARDIOGRAM

1. The cardiac cycle consists of the CONTRACTION (SYSTOLE) and RELAXATION (DIASTOLE) of both atria plus the contraction and relaxation of both ventricles followed by a brief pause.
2. The ECG (EKG) is a valuable tool for the diagnosis of abnormal cardiac rhythms and conduction patterns. An ECG is a recording of electrical changes during each cardiac cycle.
3. The normal ECG consists of a P WAVE, QRS WAVE, and a T WAVE.
4. The P WAVE represents the spread of impulses from the SA node over the atria. The QRS WAVE represents the spread of impulses through the ventricles. The T WAVE represents the repolarization of the ventricles.
5. The P-Q INTERVAL represents the conduction time from the beginning of atrial excitation to the beginning of ventricular excitation. The S-T SEGMENT represents the time between the end of the spread of impulse through the ventricles and the repolarization of the ventricles.

I. BLOOD FLOW THROUGH THE HEART

1. The movement of blood through the heart is directly related to CHANGES IN PRESSURE, which is caused by changes in the size of the chambers, which brings about the opening and closing of valves. Blood will flow through the heart from areas of HIGHER PRESSURE TO AREAS OF LOWER PRESSURE.
2. When the walls of the atria are stimulated to contract by the SA node, the subsequent decrease in size of the atria chambers causes an increase in pressure.
3. The increase in pressure forces the AV VALVES to open and atrial blood flows into the ventricles. The ventricular walls will stay relaxed until atrial contraction is complete. This causes the pressure in the ventricle to decrease.
4. When the VENTRICLE WALLS contract, the chamber size decreases and blood pressure increases to a level higher than that in the arteries.
5. The ventricular blood pushes the semilunar valves open and blood flows into the pulmonary artery and aorta. This PRESSURE OF CONTRACTION also causes the ATRIOVENTRICULAR VALVES to be pushed shut and prevent backflow of blood into the atrial chambers. Valve

cusp eversion is prevented by the chordae tendinae that attach the ventricular face of the cusp to papillary muscles in the ventricles.

6. If the cusps start to evert, the papillary muscle will contract and pull on the chordae tendinae, which will in turn pull on the ventricular face of the valve cusp and prevent eversion.

J. CARDIAC CYCLE

1. In the normal heart, the two atria contract while the two ventricles relax. The term SYSTOLE refers to the PHASE OF CONTRACTION, and the term DIASTOLE refers to the STAGE OF RELAXATION.
2. One CARDIAC CYCLE equals one complete heartbeat and consists of ATRIAL SYSTOLE and VENTRICULAR DIASTOLE occurring simultaneously, followed by ventricular systole and atrial diastole occurring simultaneously.
3. The average heart rate is 75 beats per minute and a complete cardiac cycle lasts approximately 0.8 seconds.
4. ASCULTATION is a process whereby one listens to the sounds within the body, and is usually done with a stethoscope. The sound of the heartbeat is due to turbulence in blood flow caused by the closure of the heart valves.
5. The FIRST HEART SOUND, "lubb", represents the closure of the atrioventricular valves soon after ventricular systole occurs.
6. The SECOND HEART SOUND, "dupp", represents closure of the semilunar valves close to the end of ventricular systole.

K. PHASES OF THE CARDIAC CYCLE

1. The cardiac cycle is divided into three major phases. These are RELAXATION PERIOD, VENTRICULAR FILLING, and VENTRICULAR SYSTOLE (CONTRACTION).
2. In the RELAXATION PERIOD, positioned at the end of the cardiac cycle, all four chambers are in DIASTOLE. This is the beginning of RELAXATION. The repolarization of the ventricular muscle fibers- the T wave in the EKG- initiated the relaxation period.
3. As the ventricles relax, pressure within the chambers drop and blood starts to flow from the pulmonary trunk and the aorta back towards the ventricles.
4. As this blood becomes trapped within the semilunar cusps, the semilunar valves close. Pressure continues to drop within the ventricles to a point at which it is below atrial pressure. At this point, the AV valves open and ventricular filling begins.
5. VENTRICULAR FILLING occurs just after the AV valves open. This occurs without atrial contraction.
6. Firing of the SA node results in atrial depolarization, followed by atrial contraction. This marks the end of the relaxation period.
7. Throughout the period of ventricular filling, the AV valves are open and the semilunar valves are closed.
8. VENTRICULAR SYSTOLE is initiated by the firing of the AV node, which initiates ventricular contraction. As contraction occurs, the blood is forced against the AV valves, forcing them shut.
9. As ventricular contraction continues, the pressure in the ventricles rises sharply above the pressure in the exiting vessels and the AV valves open and blood is ejected.

L. CARDIAC OUTPUT

1. The amount of blood ejected per minute from the left ventricle into the aorta, or the right ventricle into the pulmonary artery is called CARDIAC OUTPUT.
2. CARDIAC OUTPUT is determined by the volume of blood pumped by the left or right ventricle during each beat times the number of beats per minute.
3. The amount of blood ejected by the ventricle during each systole is called STOKE VOLUME and averages about 70 ml in the adult. The average heart rate is 75 beats per minute. Therefore, average CARDIAC OUTPUT is 5.25 liters per minute.
4. The length of cardiac muscle fibers influences the force of ventricular contraction. The more the cardiac muscle fibers are stretched as a chamber is filled with blood, the stronger the walls will contract to eject the blood. This relationship is known as STARLING'S LAW OF THE HEART.

M. HEART RATE

1. Heart rate is influenced by AUTONOMIC CONTROL, CHEMICALS, TEMPERATURE, EMOTIONS, GENDER, and AGE.
2. The MEDULLA of the brain contains a group of neurons called the CARDIAC CENTER. This center is comprised of two subdivisions: the CARDIOACCELERATOR CENTER (CAC) and the CARDIOINHIBITORY CENTER (CIC).
3. The CARDIOACCELERATOR CENTER gives rise to sympathetic fibers that travel to the cardiac nerves that innervate the conduction system, atria, and ventricles.
4. When the CAC is stimulated, nerve impulses travel down the sympathetic fibers, releasing NE and increasing the rate of heartbeat and strength of contraction.
5. The CIC gives rise to parasympathetic fibers that reach the heart via the VAGUS NERVE (CRANIAL NERVE X). When this center is stimulated, nerve impulses reaching the heart cause the release of AcH which decreases heartbeat and strength of contraction by slowing the SA and AV nodes.
6. The AUTONOMIC CONTROL of the heart is balanced by opposing SYMPATHETIC and PARASYMPATHETIC influences.
7. Certain chemicals have an effect on heart rate. EPINEPHRINE, released by the adrenal medulla, increases the excitability of the SA node, which in turn causes an increase in the rate and strength of contraction. Elevated potassium or sodium ions decrease heart rate and strength of contraction. An excess of CALCIUM in the blood will interfere with cardiac muscle contraction. A decrease in SERUM OXYGEN will cause heart rate to increase.
8. Increased BODY TEMPERATURE due to fever or strenuous exercise will cause the AV node to discharge impulses faster and thereby increase heart rate. Decreased body temperature decreases heart rate and strength of contraction.
9. Strong EMOTIONS such as rage, fear, anger, and anxiety increase heart rate through the general adaptation syndrome.
10. GENDER is another influencing factor. Heartbeat is slightly faster in females than in male.
11. There is a gradual decrease in heart rate as a person ages.

N. RISK FACTORS IN HEART DISEASE

1. It is estimated that 20% of the individuals that reach 60 years of age will have a myocardial infarction. Between the ages of 30 and 60 years of age, 25% of the population has the potential to be stricken.
2. RISK FACTORS are characteristics, symptoms, or signs present in a person free of disease that are statistically associated with an excessive rate of development of a disease.
3. Among the major risk factors for heart disease are high blood cholesterol level, high blood pressure, cigarette smoking, obesity, lack of regular exercise, diabetex mellitus, and a genetic predisposition.

O. COMMON DISORDERS

1. In CORONARY ARTERY DISEASE (CAD), the heart muscle does not receive an adequate amount of blood because of an interruption in the blood supply. This can be caused by atherosclerosis or coronary artery spasm.

2. ATHEROSCLEROSIS is a process in which fatty substances are deposited in the walls of the medium-sized and large arteries resulting in an impeded blood flow.
3. CORONARY ARTERY SPASM can also impede blood flow when the smooth muscle of a coronary artery undergoes contraction, especially vasoconstriction.
4. CONGESTIVE HEART FAILURE is a chronic or an acute state that results when the heart is not capable of supplying the oxygen demands of the body.

IV. TEACHING TIPS AND SUGGESTIONS

A HELPFUL HINTS

1. Dissection of the heart of a large vertebrate such as a horse is an excellent aid for clearly demonstrating the internal features of the heart.
2. In reviewing the electrocardiogram, indicate that there are three types: resting, stress, and ambulatory.

B. ESSAYS

1. Trace a drop of blood through the heart beginning with its entrance from the superior vena cava. Detail all chambers, structures, valves, and internal muscles as you progress through the cycle.
2. Impulse conduction through the conduction system of the heart generates electrical charges that may be detected on the surfaces of the body. What is a recording of these charges called? What is a deflection wave? What is the significance of each of the following: P wave, P-R interval, QRS wave, S-T segment, and T wave?

C. TOPIC FOR DISCUSSION

1. Discuss the ethical, moral, and legal ramifications of heart transplants. What are some reasons for donating the body organs of oneself or a relative? What are some reasons for not donating?

V. AUDIOVISUAL MATERIALS

A. OVERHEAD TRANSPARENCIES

1. PAP Transparency Set (Trs. 20.1a; 20.2a&b; 20.3a,c&e; 20.4a,b&c; 20.5-20.7; 30.8; 30.9; 30.22 & 30.23).
2. Human Blood Vascular System, Parts 2 & 3 (CARO).
3. Heart and Pulse Rate (CARO).

B. VIDEOCASSETTES

1. Two Hearts That Beat As One (26 min.; FHS).
2. The Physiology of Exercise (15 min.; FHS).
3. Heart Attack (29 min.; 1983; KSU).
4. Child Heart Surgery (10 min.; C; Sd; 1989; PLP).
5. Circulation (29 min.; C; Sd; 1978; GA).
6. Heart Attack: The Unrelenting Killer (29 min.; C; Sd; 1983; KSU).
7. Heart Disease (19 min.; C; Sd; 1990; FHS).
8. Heart Dissection and Anatomy (14 min.; C; Sd; 1983; KSU).
9. Heart Surgery (10 min.; C; Sd; 1990; FHS).
10. The Heart (29 min.; C; Ds; 1978; IM).
11. The Heart and the Circulatory System (16 min.; C; Sd; 1988; GA).

C. FILMS: 16 MM

1. Heart (6 min.; 1974; McG/KSU).
2. I Am Joe's Heart (26 min.; 1971; PYR/KSU).
3. The Work of the Heart (19 min.; 1967; EBEC/KSU).
4. The Heart and Circulatory System (15 min.; 1975; EBEC/KSU).
5. CPR: To Save A Life (14 min.; 1977; EBEC/KSU).
6. The Transplant Experience (50 min.; 1976; TLF/KSU).
7. A Change of Heart (15 min.; 1984; PYR/KSU).
8. Two Original Open-Heart Surgeries (EBEC).
9. Closed Chest Heart Massage (11 min.; KSU).

D. TRANSPARENCIES: 35 MM (2X2)

1. PAP Slide Set (Slides 83-90).
2. AHA Slide Set.
3. Visual Approach to Histology: Cardiovascular System (21 Sides; FAD).
4. Circulatory System and Its Function (20 Slides; EIL).

E. COMPUTER SOFTWARE

1. Heart Probe (Apple; SC-175026; PLP).
2. Cardiovascular System (Apple II; SC-182011; IBM; SC-182012; PLP).
3. The Heart Simulator (Apple; SC-920021; PLP).
4. Heart Disease (Apple; SC-390819; PLP).
5. Body Language: Cardiovascular System (Apple II Series; ESP).
6. Cardiac Muscle Mechanics (IBM; PC; QUEUE).
7. Cardiovascular Fitness Lab (Apple IIs; IBM Pc; QUEUE).
8. CPR (IBM; PLP).
9. The Body Electric (Apple IIs; QUEUE; PLP).
10. Heart Disease (Apple; PLP).

CHAPTER 16

THE CARDIOVASCULAR SYSTEM: BLOOD VESSELS

CHAPTER AT A GLANCE

- ARTERIES
- ARTERIOLES
- CAPILLARIES
- VENULES
- VEINS
- BLOOD RESERVOIRS
- PHYSIOLOGY OF CIRCULATION
    - *Blood Flow*
    - *Blood Pressure*
    - *Resistance*
- FACTORS THAT AFFECT ARTERIAL BLOOD PRESSURE
    - *Cardiac Output (CO)*
    - *Blood Volume*
    - *Peripheral Resistance*
- HOMEOSTASIS OF BLOOD PRESSURE REGULATION
    - *Vasomotor Center*
    - *Baroreceptors*
    - *Chemoreceptors*
    - *Regulation by Higher Brain Centers*
    - *Hormones*
    - *Autoregulation*
- CAPILLARY EXCHANGE
- FACTORS THAT AID IN VENOUS RETURN
    - *Pumping Action of the Heart*
    - *Velocity of Blood Flow*
    - *Skeletal Muscle Contractions and Valves*
    - *Breathing*
- CHECKING CIRCULATION
    - *Pulse*
    - *Measurement of Blood Pressure (BP)*
- SHOCK AND HOMEOSTASIS
- CIRCULATORY ROUTES
    - *Systemic Circulation*
    - *Pulmonary Circulation*
    - *Cerebral Circulation*
    - *Hepatic Portal Circulation*
    - *Fetal Circulation*
- COMMON DISORDERS
- MEDICAL TERMINOLOGY AND CONDITIONS
- WELLNESS FOCUS: HYPERTENSION PREVENTION

I. CHAPTER SYNOPSIS

The main histological and physiological features of blood vessels are presented in this chapter. The concept of blood reservoirs is explained and the physiology of circulation is discussed with emphasis on blood flow, factors that affect arterial blood pressure, control of blood pressure, and factors that aid in venous return. There is extensive discussion of circulatory shock and the relationship of its stages to homeostasis. There are descriptions of the measurement of pulse and blood pressure as a means of checking circulation. The architecture and functions of systemic, pulmonary, hepatic portal, and fetal circulatory routes are carefully considered. The effects of exercise and aging on the cardiovascular system are also covered. The chapter concludes with a list of common disorders, medical terms, and conditions associated with the cardiovascular system.

II. LEARNING GOALS/STUDENT OBJECTIVES

1. Describe the factors that affect blood pressure.
2. Describe how blood pressure is regulated.
3. Discuss how materials are exchanged between blood cells and body cells.
4. Describe how blood returns to the heart.
5. Define pulse and blood pressure and describe how they are measured.
6. Compare the major routes that blood takes through various regions of the body.

III. SAMPLE LECTURE OUTLINE

A. ARTERIES

1. ARTERIES carry blood away from the heart and are constructed of three layers of tissue and a hollow core called the lumen. It is through the lumen that blood flows.
2. The INNERMOST LAYER of an artery wall is the ENDOTHELIUM, which is made of simple squamous epithelial and elastic tissues. The MIDDLE LAYER consists of SMOOTH MUSCLE and ELASTIC CONNECTIVE TISSUE, and the OUTER LAYER is principally composed of ELASTIC and COLLAGENOUS fibers.
3. As a result of the structure of the middle layer, arteries have two major properties: elasticity and contractility.
4. The ELASTICITY allows arteries such as the aorta and pulmonary arteries to stretch and recoil, dependent upon blood pressure.
5. The CONTRACTILITY of an artery arises from its smooth muscle layers, which are supplied by the sympathetic branch of the autonomic nervous system. Sympathetic stimulation of an artery causes vasoconstriction, a decrease in the diameter. The inhibition of sympathetic stimulation leads to vasodilation, an increase in the diameter.

B. ARTERIOLES

1. ARTERIOLES are small arteries that function to deliver blood to the capillaries.
2. The closer arterioles are to the capillaries, the smaller they are and fewer fibers they contain.

3. Arterioles function in the REGULATION OF BLOOD FLOW from arteries to capillaries, and the regulation of blood flow by vasoconstriction and vasodilation.

C. CAPILLARIES

1. CAPILLARIES are microscopic vessels that connect the arterioles to venules.
2. The primary function of the capillaries is to permit the exchange of OXYGEN, NUTRIENTS, and WASTES between the blood and the tissue cells.
3. Capillaries differ structurally from arteries and arterioles in that they are comprised of a single layer of ENDOTHELIAL cells and allow the passage of materials between their cell junctions.
4. Some capillaries, called true capillaries, contain a precapillary sphincter or ring of smooth muscle at the junction of the arteriole and capillary, which serves to regulate the blood flow through the capillary.

D. VENULES

1. VENULES are formed by the union of several capillaries and serve to collect blood from these capillaries and drain it into the veins.
2. Venules are structurally similar to arterioles but contain less muscle and lead directly into veins.

E. VEINS

1. VEINS are structurally similar to arteries but their outer layer is thicker, their middle layer thinner, and the inner layer of endothelium may fold inward to form valves.
2. Blood leaving the capillaries and moving into the veins loses pressure, and therefore will flow more slowing and evenly in the vein.
3. The low pressure in veins can cause the backflow of blood, especially when the forces of gravity are working. The presence of VALVES in the veins, especially in the extremities, prevents backflow.

F. BLOOD RESERVOIRS

1. The volume of blood in various parts of the cardiovascular system varies considerably.
2. SYSTEMIC VEINS are called BLOOD RESERVOIRS. They store blood and through venous vasoconstriction can move blood to other parts of the body as needed.
3. In cases of hemorrhage, when blood pressure and volume decreases, vasoconstriction of the veins in venous reservoirs helps to compensate for the blood loss.
4. The principal reservoirs are associated with the veins of the abdominal organs, the liver and spleen, and the skin.

G. PHYSIOLOGY OF CIRCULATION

1. BLOOD FLOW refers to the amount of blood that passes through a blood vessel in a given period of time.
2. Blood flows, or circulates, through two major sets of blood vessels: to the lungs where carbon dioxide is exchanged for oxygen (pulmonary circulation), and to the rest of the body

where the oxygenated blood is distributed and carbon dioxide is removed (systemic circulation).

3. Blood flow is determined by BLOOD PRESSURE and RESISTANCE.
4. BLOOD PRESSURE is the pressure exerted by blood on the walls of the blood vessels, and is influenced by cardiac output, blood volume and resistance.
5. Blood flows through its system of closed vessels because of different blood pressures in the various parts of the cardiovascular system.
6. Blood flow is directly proportional to blood pressure because blood always flows from a region of higher blood pressure to a region of lower blood pressure.
7. RESISTANCE refers to the opposition to blood flow that results from friction between the blood and the blood vessel walls, and is related to blood viscosity, blood vessel length, and blood vessel radius.
8. The VISCOSITY of blood is a function of the ratio of the volume of formed and dissolved components of blood to the volume of plasma. Any condition that increases viscosity also increases blood pressure.
9. The longer the blood vessel, the greater the resistance as blood flows through it.
10. The smaller the radius, the greater the resistance it offers to blood flow.

H. FACTORS THAT AFFECT ARTERIAL BLOOD PRESSURE

1. Three factors influence arterial blood pressure: CARDIAC OUTPUT, BLOOD VOLUME, and PERIPHERAL RESISTANCE.
2. CARDIAC OUTPUT is the amount of blood ejected by the left ventricle into the aorta or right ventricle into the pulmonary artery each minute. It is the principal determinant of blood pressure.
3. BLOOD PRESSURE varies directly with cardiac output. Any increase in cardiac output will increase blood pressure. Conversely, any decrease in cardiac output will decrease blood pressure.
4. Blood pressure is directly proportional to the volume of blood in the cardiovascular system. Any decrease or increase in the volume will decrease or increase blood pressure, respectively.
5. PERIPHERAL RESISTANCE is the opposition to blood flow in peripheral circulation. Arterioles control peripheral resistance, and therefore blood pressure and blood flow as well, by changing diameter. This is under AUTONOMIC CONTROL.

I. HOMEOSTASIS OF BLOOD PRESSURE REGULATION

1. In order to maintain homeostasis, blood pressure must be kept within a normal range. Both high and low blood pressure can cause serious damage to tissues and organs. A REGULATORY CENTER in the brain receives input from receptors throughout the body from higher brain centers and from numerous chemicals.
2. The VASOMOTOR CENTER is a cluster of sympathetic neurons located in the medulla which control the diameter of blood vessels, especially the arterioles of the skin and abdominal viscera.
3. The VASOMOTOR CENTER functions with other integrating centers for blood pressure control and continually sends impulses to the smooth muscles of the arteriole walls that results in a moderate state of VASOCONSTRICTION at all times. This maintains blood pressure and peripheral resistance.

4. If the number of sympathetic impulses increases, the vasomotor center brings about further vasoconstriction and increased blood pressure. Conversely, if the number of sympathetic impulses decreases, VASODILATION occurs and a lowering of blood pressure results.
5. The sympathetic division of the ANS can bring about either vasoconstriction or vasodilation by varying the number of impulses sent to the vasomotor center.
6. The vasomotor center can be modified by input from BARORECEPTORS, CHEMORECEPTORS, HIGHER BRAIN CENTERS, and CHEMICALS, all of which influence blood pressure.
7. BARORECEPTORS are neurons that are sensitive to blood pressure. They are located in the aorta, internal carotid arteries, and other large arteries in the neck and chest.
8. The BARORECEPTORS help to regulate blood pressure by sending impulses to the cardiac center to increase or decrease cardiac output.
9. CHEMORECEPTORS are neurons that are sensitive to chemicals in the blood. They are located in the two carotid bodies and in several aortic bodies.
10. CHEMORECEPTORS are sensitive to lower than normal levels of oxygen and higher than normal levels of carbon dioxide and hydrogen ions, and send impulses to the vasomotor center when these conditions exits.
11. In response, the vasomotor center increases sympathetic stimulation to the arterioles, which results in vasoconstriction and an increase in blood pressure.
12. HIGHER BRAIN CENTERS, such as the cerebral cortex, influence blood pressure in response to strong emotions such as anger and sexual excitement. In response to these emotions, the cerebral cortex relays impulses to the hypothalamus, which passes the impulses onto the vasomotor center.
13. The vasomotor center sends impulses to the arterioles to initiate vasoconstriction and an increase in blood pressure, and to the adrenal medulla which will release the vasoconstrictors epinephrine and norepinephrine.
14. Numerous HORMONES also affect blood pressure by causing vasoconstriction. Among the chemicals are EPINEPHRINE, NE, ALDOSTERONE, ANGIOTENSIN, and ADH. Other chemicals, such as histamine, kinins, and alcohol, inhibit the vasomotor center, evoke vasodilation, and lower blood pressure.
15. AUTOREGULATION, or a local, autonomic adjustment in blood flow in a given region of the body, occurs in response to the particular needs of a certain tissue. In most instances, oxygen is the indirect stimulation for autoregulation.
16. In response to low oxygen supplies, cells produce vasodilator substances such as potassium and hydrogen ions, carbon dioxide, lactic acid and/or adenosine, which cause local arterioles to vasodilate. This results in an increase in blood flow to the tissue, restoring oxygen levels.
17. A number of factors help blood return through the veins to the heart by increasing the magnitude of the pressure gradient between the vein and the right atrium. These factors include the velocity of blood flow, skeletal muscle contraction, valves in veins, and breathing.

J. CAPILLARY EXCHANGE

1. The VELOCITY of blood flow in the capillaries is the slowest in the cardiovascular system. This allows for the exchange of materials between the blood and body tissues.
2. Blood does not flow steadily through the capillaries but rather intermittently because of the contraction and relaxation of the smooth muscle fibers and METARTERIOLES and the PRECAPILLARY SPHINCTERS of true capillaries. This intermittent contraction and relaxation is called VASOMOTION.

3. The movement and exchange of materials occurs by FILTRATION.

K. FACTORS THAT AFFECT VENOUS RETURN

1. Several factors allow for venous return to occur successfully. These are:
    - pumping action of the heart
    - velocity of blood flow
    - skeletal muscle contractions
    - valves in veins
    - breathing

L. CHECKING CIRCULATION

1. PULSE is the alternating expansion and elastic recoiling of the wall of an artery with each systole and diastole of the left ventricle.
2. Pulse may be felt in any large artery that lies near the surface or over hard tissues, and is stronger closer to the heart.
3. The normal pulse is between 70 to 90 beats per min. in the resting state.
4. Blood pressure is the pressure exerted by blood on the wall of an artery when the left ventricle undergoes systole and the pressure remaining in the arteries when the ventricle is in diastole.
5. SYSTOLIC BLOOD PRESSURE is the force of the blood recorded during ventricular contraction.
6. DIASTOLIC BLOOD PRESSURE is the force of the blood recorded during ventricular relaxation.

M. MEASUREMENT OF BLOOD PRESSURE

1. BLOOD PRESSURE refers to the pressure in the arteries exerted by the left ventricle when it undergoes systole and the pressure remaining in the arteries when the ventricle is in diastole.
2. Blood pressure is usually taken in the LEFT BRACHIAL ARTERY on the arm and is measured by a SPHYGMOMANOMETER.
3. After the pressure cuff is inflated, the artery is compressed so that the blood stops to flow.
4. As the cuff is deflated, the artery opens, and a spurt of blood passes through. This results in the first sound heard by the stethoscope and corresponds to the SYSTOLIC BLOOD PRESSURE (SBP). This is the force with which the blood is pushing against the arterial walls during ventricular contraction.
5. The pressure recorded when the sounds suddenly become faint is called the DIASTOLIC BLOOD PRESSURE (DBP) and measures the force of blood remaining in the arteries during ventricular relaxation.

N. SHOCK AND HOMEOSTASIS

1. Shock occurs when the cardiovascular system cannot deliver sufficient oxygen and nutrients to meet the needs of the body. The underlying cause is insufficient cardiac output.
2. Inadequate cardiac output results in cellular membrane dysfunctions, abnormal cell metabolism, and possibly cellular death.

3. The causes of inadequate cardiac output may be hemorrhage, burns, or dehydration due to excessive vomiting, diarrhea, or sweating.

O. CIRCULATORY ROUTES

1. The blood vessels are organized into definite routes to circulate blood throughout the body.
2. The largest circulatory route is SYSTEMIC circulation.
3. Two of the several subdivisions of systemic circulation are CORONARY CIRCULATION and HEPATIC PORTAL CIRCULATION. Other routes include PULMONARY, CEREBRAL, and FETAL CIRCULATION.
4. The SYSTEMIC ROUTE takes oxygenated blood from the left ventricle through the aorta to all parts of the body, including the tissue of the lung, but does not supply the alveoli of the lung for gaseous exchange. Systemic circulation returns blood to the right atrium.
5. The AORTA is divided into the ascending aorta, aortic arch, and descending aorta. Each section of the aorta gives off branches of arteries that will supply the whole body.
6. Blood is returned to the heart through systemic veins. All veins of systemic circulation flow into either the SUPERIOR or INFERIOR VENA CAVAE, or into the CORONARY SINUS. All three empty into the RIGHT ATRIUM.
7. PULMONARY CIRCULATION takes deoxygenated blood from the right ventricle to the lungs and returns oxygenated blood from the lung alveoli to the left atrium. It allows blood to be oxygenated for systemic circulation.
8. HEPATIC PORTAL CIRCULATION collects blood from the veins of the pancreas, spleen, stomach, intestines, and gall bladder, and directs it to the hepatic portal vein of the liver before it returns it to the heart.
9. The HEPATIC PORTAL ROUTE enables the liver to store nutrients, detoxify harmful substances in the blood, and phagocytize any bacteria which may be present.
10. The fetus derives its oxygen and nutrients and eliminates its wastes and carbon dioxide through the maternal blood supply by means of a structure called the placenta. FETAL CIRCULATION involves the exchange of materials between the fetus and the placenta.
11. CEREBRAL CIRCULATION is accomplished through an arrangement of blood vessels at the base of the brain called the cerebral arterial circle or CIRCLE OF WILLIS.
12. The CIRCLE OF WILLIS is formed by the union of the internal carotid artery and its branches, the anterior cerebral arteries, and the basilar artery and its branches, and the posterior cerebral artery.

IV. TEACHING TIPS AND SUGGESTIONS

A. HELPFUL HINTS

1. If dissection of an organism is possible, the major routes of circulation and the major blood vessels should be exposed.
2. Ask the students to describe blood flow through any number of circulatory routes.
3. Give the student a particular organ and ask which vessel is its supplying vessel and which vessel is its emptying vessel. This is best illustrated with the hepatic and fetal circulatory routes.

4. Inform the student that vessels are not named by the type of blood they are carrying, but rather their direction. Provide examples such as the pulmonary arteries and veins and the umbilical vein.

B. ESSAY QUESTIONS

1. Detail the flow of blood from the superior vena cava to its exit at the aorta. Include all organs and physiological functions.
2. Detail the flow of blood through the hepatic portal route. Why are materials re-routed from the heart to the liver? Is there a practical advantage to this?
3. Assume that a large blood clot has lodged in the left common iliac artery. List at least five other arteries that would have a reduced blood supply because of this clot. What parts of the body would be affected?

C. TOPIC FOR DISCUSSION

1. Discuss the principal differences between adult and fetal circulations relative to the type of blood carried in the vessels. Discuss why a fetus is able to carry mostly mixed blood in the major vessel.

V. AUDIOVISUAL MATERIALS

A. OVERHEAD TRANSPARENCIES

1. PAP Transparency Set (Trs. 21.1a-c, 21.2 a&b, 21.4a, 21.8-21.10, 21.11a&b, 21.12-21.17, 21.19, 21.20a&b, 21.22a-c, 21.24a&b, 21.25, 21.27, 21.29a&b, 21.30a, 21.31a & 21.32).
2. Fetal Circulation (CARO).
3. Human Blood Vascular System I (CARO).
4. Blood Circulation, Arteries, and Veins (CARO).
5. Mammalian Circulatory System (K&E).

B. VIDEOCASSETTES

1. Life Under Pressure (26 min.; FHS).
2. The Silent Killer: Hypertension (12 min.; 1983; KSU).
3. Your Blood Pressure Is Showing (29 min.; 1983; KSU).
4. Blood: The Vital Humor (30 min.; HR).
5. Hypertension: Your Blood Pressure is Showing (29 min.; C; Sd; 1983; KSU).

C. FILMS: 16 MM

1. The Human Body: Circulatory System (16 min.; 1980; COR/KSU).
2. The Heart and Circulation (15 min.; 1975; EBEC/KSU).
3. Hemo the Magnificent (59 min.; WAVE).
4. Arterial Blood Pressure Regulation (18 min.; 1970; MG/KSU/UIFC).
5. Arteries and Veins (14 min.; KSU).

D. TRANSPARENCIES: 35 MM (2X2)

1. PAP Slide Set (Slides 91-100).
2. AHA Slide Set.
3. Visual Approach to Histology: Cardiovascular System (21 Slides; FAD).
4. Circulatory System (Slides 83-106; McG).
5. Circulating Blood, Blood Vessels, and Bone Marrow (29 Slides; CARO).

E. COMPUTER SOFTWARE

1. Cardiovascular System (Apple II; SC-182011; IBM; SC-182012; PLP).
2. Heart Disease (Apple: PLP).
3. Body Language: Cardiovascular System (Apple II Series; ESP).
4. Biomedical Software: A&P Concepts (Apple; PLP).
5. Flash: Blood Vessels (Apple; IBM; PLP).
6. Graphic Human Anatomy and Physiology Tutor: Blood Vessels (IBM; PLP).
7. Understanding Cardiovascular Function (IBM; PLP).

CHAPTER 17

THE LYMPHATIC SYSTEM AND IMMUNITY

CHAPTER AT A GLANCE

- FUNCTIONS
- LYMPH AND INTERSTITIAL FLUID
- LYMPHATIC CAPILLARIES AND LYMPHATIC VESSELS
- LYMPHATIC TISSUE
    - *Lymph Nodes*
    - *Tonsils*
    - *Spleen*
    - *Thymus Gland*
- LYMPH CIRCULATION
- NONSPECIFIC RESISTANCE TO DISEASE
    - *Skin and Mucous Membranes*
        - *Mechanical Factors*
        - *Chemical Factors*
    - *Antimicrobial Substances*
        - *Interferon (IFN's)*
        - *Complement*
    - *Natural Killer (NK) Cells*
    - *Phagocytosis*
        - *Kinds of Phagocytes*
        - *Mechanism*
    - *Inflammation*
        - *Symptoms*
        - *Stages*
    - *Fever*
- IMMUNITY (SPECIFIC RESISTANCE TO DISEASE)
    - *Antigens*
    - *Antibodies*
    - *Cell-mediated Antibody-mediated*
    - *Antibody-mediated*
    - *Formation of T-cells and B-cells*
    - *T-cells and Cell-mediated Immunity*
    - *B-cells and Antibody-mediated Immunity*
    - *Actions and Antibodies*
    - *The Skin and Immunity*
    - *Immunology and Cancer*
- COMMON DISORDERS
- MEDICAL TERMINOLOGY AND CONDITIONS
- WELLNESS FOCUS: MIND AND IMMUNITY

I. CHAPTER SYNOPSIS

The lymphatic system is composed of lymph, lymphatic vessels, lymph nodes, and three organs: the tonsils, the thymus gland, and the spleen. Attention is given to the histology of lymph vessels, lymph nodes, the plan of lymph circulation, and the structure and function of the lymphatic organs. Other principal areas discussed are the various kinds of nonspecific resistance to disease, the role of lymphatic tissue in antibody production, the skin and immunity, monoclonal antibodies, immunology, and cancer. Disorders that are discussed include an up-to-date section on Acquired Immune Deficiency Syndrome (AIDS), autoimmune diseases, severe combined immunodeficiency, hypersensitivity tissue rejection, and Hodgkin's disease. The chapter concludes with a list of medical terms, and conditions associated with the lymphatic system and immunity.

II. LEARNING GOALS/STUDENT OBJECTIVES

1. Explain how lymph and interstitial fluid are related.
2. Compare the structure and functions of the various types of lymphatic tissue.
3. Describe how lymph circulates throughout the body.
4. Explain how the skin and mucous membranes protect the body against disease.
5. Describe how various microbial substances protect the body against disease.
6. Define phagocytosis and explain how it occurs.
7. Describe how inflammation occurs and why it is important.
8. Define immunity and compare it with nonspecific resistance to disease.
9. Explain the relationship between an antigen and an antibody.
10. Compare the functions of cell-mediated and antibody-mediated immunity.

III. SAMPLE LECTURE OUTLINE

A. FUNCTIONS

1. The lymphatic system consists of a fluid called LYMPH, LYMPHATIC VESSELS, LYMPH NODES, and three organs: the TONSILS, the THYMUS GLAND, and the SPLEEN.
2. The primary tissue of the lymphatic system is LYMPHATIC TISSUE, a specialized type of loose reticular connective tissue.
3. Lymph tissue is found in various forms in the body. If it is not enclosed by a membrane it is referred to as diffuse lymphatic tissue.
4. Diffuse lymphatic tissue is found in the mucous membranes of the gastrointestinal tract, respiratory passageways, urinary, and reproductive tracts.
5. LYMPHATIC NODES are non-encapsulated masses of lymphatic tissue which are present in the gastrointestinal tract, urinary and reproductive tracts, and comprise the tonsils.
6. The LYMPHATIC ORGANS are the LYMPHATIC NODES, SPLEEN, THYMUS GLAND, and BONE MARROW.
7. The lymphatic system functions to drain tissue spaces of protein-containing fluid that escaped from blood capillaries and return the protein to the cardiovascular system, transport fats from the gastrointestinal tract to the blood, and protect the body from foreign cells, microbes, and cancer cells.

B. LYMPH AND INTERSTITIAL FLUID

1. Fluid found immediately around cells is called INTERSTITIAL FLUID. Interstitial fluid which flows in defined vessels is called LYMPH.
2. Both interstitial fluid and lymph are similar in composition to plasma, but contain less protein.
3. Approximately 3 liters per day of fluid seeps from the blood into the tissue. This fluid and protein must be returned to the cardiovascular system to maintain normal blood volume and homeostasis.

C. LYMPHATIC VESSELS

1. LYMPHATIC VESSELS originate as LYMPH CAPILLARIES, (microscopic vessels between cells), occurring singly or in networks. They are found throughout the body with the exception of avascular tissue, the central nervous system, and bone marrow.
2. Lymphatic capillaries merge to form larger lymphatic vessels, which in turn, converge to form lymphatic ducts. These vessels will drain the lymph into the left and right subclavian veins, respectively.
3. LYMPHATIC VESSELS have thinner walls and more valves than veins. The LEFT LYMPHATIC DUCT is known as the THORACIC DUCT and drains into the LEFT SUBCLAVIAN VEIN. The RIGHT LYMPHATIC DUCT drains into the RIGHT SUBCLAVIAN VEIN.

D. LYMPHATIC TISSUE

1. LYMPH NODES are oval structures located along the length of lymphatic vessels which are scattered throughout the body.
2. Lymph enters the nodes through AFFERENT LYMPHATIC VESSELS and exits through EFFERENT LYMPHATIC VESSELS. While passing through the nodes it is filtered to remove damaged cells and microorganisms. Lymph nodes also produce lymphocytes.
3. TONSILS are multiple aggregations of large lymphatic nodules embedded in mucous membranes. They include the PHARYNGEAL, PALATINE, and LINGUAL TONSILS.
4. The tonsils are strategically situated to protect against antigens that enter the oropharynx and nasopharynx and function to produce lymphocytes and antibodies.
5. The SPLEEN is the largest mass of lymphatic tissue in the body and is found in the left hypochondriac region between the fundus of the stomach and the diaphragm.
6. The spleen functions to produce B-LYMPHOCYTES and ANTIBODIES, and to phagocytize bacteria and worn-out or damaged red blood cells and platelets. The spleen is also a STORAGE ORGAN for blood.
7. The THYMUS functions in immunity by the production and distribution of T-LYMPHOCYTES (T-CELLS). It is located in the superior anterior mediastinum, posterior to the sternum and medial to the lungs.

E. LYMPH CIRCULATION

1. When plasma is filtered by blood capillaries, it passes into the interstitial spaces and becomes INTERSTITIAL FLUID. When this fluid passes into the lymphatic capillaries, it is called LYMPH.

2. LYMPH passes from the interstitial fluid to lymph capillaries to lymphatic vessels to the thoracic duct or right lymphatic duct to the subclavian veins.
3. Lymph flows primarily as a result of skeletal muscle contractions and respiratory movements. It is also aided by valves located within the lymphatic vessels.

F. NONSPECIFIC RESISTANCE TO DISEASE

1. The ability to ward off disease using a number of defenses is called RESISTANCE. The lack of resistance is called SUSCEPTIBILITY.
2. NONSPECIFIC RESISTANCE is inherited and refers to a wide variety of body responses against a wide range of pathogens, toxins, or disease-producing organisms. Specific resistance or immunity is the ability to produce antibodies against specific pathogens.
3. Nonspecific resistance includes a number of MECHANICAL and CHEMICAL FACTORS.
4. MECHANICAL FACTORS include INTACT SKIN, MUCOUS MEMBRANES, and LACRIMAL APPARATI, the PRESENCE of SALIVA, MUCUS, CILIA, the EPIGLOTTIS, and the FLOW of URINE.
5. CHEMICAL FACTORS include ANTIMICROBIAL SUBSTANCES secreted by the skin and acid released by the stomach.
6. ANTIMICROBIAL SUBSTANCES in the blood and tissues such as INTERFERON, COMPLEMENT, and PROPERDIN, work against colonization by viruses and bacteria.
7. Phagocytosis is a nonspecific mechanism by which microorganisms and foreign particles are ingested by white blood cells such as neutrophils, eosinophils, and wandering macrophages.
8. The INFLAMMATORY RESPONSE serves a protective and defensive role by eliminating microbes or foreign substances from the site of injury, preventing their spread to other organs, and preparing the site for tissue repair. It is an attempt to restore tissue homeostasis.
9. INFLAMMATION occurs when cells are damaged by microbes, physical agents, or chemical agents. The symptoms of inflammation include redness, pain, heat, and swelling.
10. The INFLAMMATORY RESPONSE may include vasodilation, increased permeability of blood vessels, fibrin formation, phagocyte migration, and pus formation.

G. IMMUNITY (SPECIFIC RESISTANCE TO DISEASE)

1. SPECIFIC RESISTANCE to disease involves the production of a specific type of cell (LYMPHOCYTE) or a specific type of molecule (ANTIBODY) to destroy a particular antigen, and is called IMMUNITY.
2. ACQUIRED IMMUNITY refers to immunity gained as a result of contact with a specific antigen. It may be obtained ACTIVELY or PASSIVELY, or by artificial means. The types of acquired immunity are: NATURALLY ACQUIRED ACTIVE IMMUNITY, NATURALLY ACQUIRED PASSIVE IMMUNITY, ARTIFICIALLY ACQUIRED ACTIVE IMMUNITY, and ARTIFICIALLY ACQUIRED PASSIVE IMMUNITY.
3. NATURALLY ACQUIRED ACTIVE IMMUNITY is obtained when the immune system comes in contact with microbes in the course of daily living and the body responds by producing T-cells and/or antibodies.
4. NATURALLY ACQUIRED PASSIVE IMMUNITY involves the transfer of antibodies from an immunized donor to a non-immunized recipient, as in antibodies crossing the placental barrier.
5. ARTIFICIALLY ACQUIRED ACTIVE IMMUNITY results from vaccination in which the individual is given the antigen that is antigenic but not pathogenic.

6. ARTIFICIALLY ACQUIRED PASSIVE IMMUNITY is gained by an injection of antibodies from outside sources.
7. ANTIGENS (Ags), or IMMUNOGENS, are chemical substances that stimulate the production of antibodies.
8. ANTIBODIES (Abs) are proteins produced in response to antigens. Antibodies respond to antigenic determinant sites on the surface of the antigen. Most antigens have several antigenic determinant sites.

H. CELL-MEDIATED AND ANTIBODY-MEDIATED IMMUNITY

1. CELL-MEDIATED IMMUNITY refers to the destruction of antigens by T-cells. Antibody-mediated immunity refers to the destruction of antigens by antibodies.
2. CELL-MEDIATED IMMUNITY is particularly effective against fungi, parasite, intracellular viral infections, cancer cells, and foreign tissue transplants. Antibody-mediated immunity is most effective against viral and bacterial infections.
3. T-CELLS, or T-LYMPHOCYTES, are responsible for cellular immunity and are processed in the thymus gland. B-CELLS, or B-LYMPHOCYTES, provide antibody-mediated immunity and are processed in the bone marrow, fetal liver tissue and spleen, and gut-associated lymphoid tissue.
4. MACROPHAGES process and present antigens to T-cells and B-cells and secrete INTERLEUKIN-1 (IL-1), which induces the proliferation of T-cells and B-cells.
5. T-cells consist of several subpopulations: CYTOTOXIC T-CELLS, HELPER T-CELLS, AMPLIFIER-T CELLS, MEMORY T-CELLS, DELAYED HYPERSENSITIVITY T-CELLS, and SUPPRESSOR T-CELLS.
6. CYTOTOXIC T-CELLS migrate to the site of invasion from the lymphoid tissue and secrete LYMPHOTOXIN (LT) that destroys the antigens directly by lysis, and secrete other LYMPHOKINES that destroy antigens indirectly.
7. HELPER T-CELLS cooperate with B-cells to help amplify antibody production and secrete Interleukin-2 (IL-2), which stimulates the proliferation of cytotoxic T-cells.
8. AMPLIFIER T-CELLS stimulate helper and suppressor T-cells and plasma cells to greater levels of activity.
9. MEMORY T-CELLS recognize antigens to which they have been sensitized, again at a later date.
10. DELAYED HYPERSENSITIVITY T-CELLS produce lymphokines and are important in hypersensitivity responses.
11. SUPPRESSOR T-CELLS inhibit the secretion of injurious substances by cytotoxic T-cells and the production of antibodies by plasma cells.
12. B-CELLS develop into antibody-producing plasma cells under the influence of thymic hormones and IL-2. Memory B-cells recognize the original invading antigen at a time of subsequent exposure.
13. Antibodies produced by B-cells enter circulation and form antigen-antibody complexes with foreign antigens. These antibodies activate COMPLEMENT enzymes for attack and fix the complement to the surface of the antigen.
14. The secondary response provides the basis for immunization against certain diseases and is usually swifter and of a greater magnitude than the original antibody response.

I. DISORDERS

1. ACQUIRED IMMUNE DEFICIENCY SYNDROME (AIDS) lowers the body's immunity by decreasing the number of helper T-cells and reversing the ratio of helper T-cells to suppressor T-cells. Persons with AIDS frequently develop Kaposi's sarcoma, a type of skin cancer, and Pneumocystis carinii pneumonia, caused by a fungus which rarely affects healthy people.
2. AUTOIMMUNE DISEASES result when the body does not recognize "self" antigens and produces antibodies against them.
3. Severe combined immunodeficiency is an immunodeficiency disease in which both B-cells and T-cells are missing or inactive in providing immunity.
4. Hypersensitivity is overreacting to an antigen. Localized anaphylactic reactions include hay fever, asthma, eczema, and hives.

IV. TEACHING TIPS AND SUGGESTIONS

A. HELPFUL HINTS

1. A discussion of the effects, modes of transmission, and prevention of AIDS is of paramount importance.
2. A discussion detailing the role of immunoglobulins is useful.
3. Comparative transparencies of systemic cardiovascular circuits and the lymphatic circuit are helpful in illustrating the interrelationship between the two.

B. ESSAYS

1. Draw a diagram to indicate the role of the thoracic duct and the right lymphatic duct in draining lymph from different regions of the body. Use arrows to indicate the direction of flow.
2. Give a complete explanation of mechanisms used to ward off disease. Compare and contrast nonspecific and specific defenses.

C. TOPICS FOR DISCUSSION

1. Discuss the importance of secondary responses of the body to antigens.
2. Discuss the role that the HIV virus has played on the human society.

V. AUDIOVISUAL MATERIALS

A. OVERHEAD TRANSPARENCIES

1. PAP Transparency Set (Trs. 22.1, 22.2a&b, 22.3-22-5, 22.9, 22.10a, 22.12, 22.14, 22.15, 22.16-22.18 & 22.20).
2. Disease and Health (12 Overheads; CARO).
3. Human Lymphatic System (CARO).

B. VIDEOCASSETTES

1. Internal Defenses (26 min.; FHS).
2. The Body Against Disease (HRM).
3. Viruses: The Mysterious Enemy (HRM).
4. AIDS: Facts and Fears, Crisis and Controversy (EIL).
5. AIDS: What are the Risks? (HRM).
6. AIDS: Can I Get It? (48 min.; 1988; EIL).
7. The Gift of Life (30 min.; 1985; CARSL/KSU).
8. Immune Response (30 min.; CRM).
9. AIDS (12 min.; C; Sd; 1991; PLP).
10. AIDS: Can I Get It? (48 min.; C; Sd; 1988; EIL).
11. AIDS: Changing Lifestyles (15 min.; C; Sd; 1991; GAF).
12. AIDS: Everyday Precautions for Healthcare Workers (90 min.; C; Sd; 1990; GA).
13. AIDS: Facts and Fears- Crisis and Controversy (56 min.; C; Sd; 1990; GA/EI).
14. AIDS: Nobody is Immune (15 min.; C; Sd; 1990; GA).
15. AIDS: On The Trail of a Killer (52 min.; C; Sd; 1990; FHS).
16. AIDS: Our Worst Fears (57 min.; C; Sd; 1990; FHS).
17. AIDS: The Global Impact (15 min.; C; SD; 1990; GA).
18. AIDS: What are the Risks? (1991; HRM).
19. AIDS: What Everyone Needs to Know (19 min.; C; Sd; 1990; KSU).
20. Immunodeficiency: A Disease of Life (19 min.; C; Sd; 1989; IM).
21. Measles (10 min.; C; Sd; 1989; PLP).
22. Ryan White Talks to Kids About Aids (28 min.; C; Sd; 1990; FHS).
23. The Biology of Viruses (19 min.; C; Sd; PLP).
24. Living with Cancer: Hodgkin's Disease (26 min.; C; Sd; 1990; FHS).
25. The Microbiology of AIDS (10 min.; C; Sd; 1990; FHS).

C. FILMS: 16 MM

1. Blueprints in the Bloodstream (Nova series) (49 min.; 1978; TLF/KSU/UIFC).
2. Body Defenses Against Disease (14 min.; 1978; EBEC/KSU).
3. Antibody Diversity and Immunoregulation (24 min.; 1977; KSU).
4. The Transplant Experience (50 min.; 1976; KSU).
5. Mechanisms of Defense (26 min.; FHS).

D. TRANSPARENCIES: 35 mm (2x2)

1. PAP Slide Set (Slides 101-105).
2. AHA Slide Set.
3. Visual Approach to Histology: Lymphatic System (18 Slides; FAD).
4. Antigens and Immunogens (55 Slides; EIL).
5. Antigen-Antibody Reactions (98 Slides; EIL).
6. Immunologic Deficiency States (110 Slides; EIL).
7. The Immune Response (10 Slides; EIL).

E. Computer Software

1. Body Defenses (Apple; SC-381047; IBM; SC-381048; PLP).
2. Understanding AIDS (Apple; SC-176015; IBM; SC-176016; PLP).
3. AIDS: What You Should Know (Apple; PLP).
4. Understanding AIDS (Apple; IBM; PLP).
5. Choices: AIDS (IBM; PLP).
6. Allergist (IBM; PLP).

CHAPTER 18

THE RESPIRATORY SYSTEM

CHAPTER AT A GLANCE

- ORGANS
- NOSE
    - *Anatomy*
    - *Functions*
- PHARYNX
- LARYNX
    - *Anatomy*
    - *Voice Production*
- TRACHEA
- BRONCHI
- LUNGS
    - *Gross Anatomy*
    - *Lobes and Fissures*
    - *Lobules*
    - *Alveolar-Capillary*
    - *Respiratory Membrane*
    - *Blood Supply*
- RESPIRATION
    - *Pulmonary Ventilation*
        - *Inspiration*
        - *Expiration*
        - *Collapsed Lung*
- PULMONARY AIR VOLUMES AND CAPACITIES
    - *Pulmonary Volumes*
    - *Pulmonary Capacities*
- EXCHANGE OF RESPIRATORY GASES
- EXTERNAL (PULMONARY) RESPIRATION
- INTERNAL (TISSUE) RESPIRATION
- TRANSPORT OF RESPIRATORY GASES
    - *Oxygen*
    - *Carbon Dioxide*
- CONTROL OF RESPIRATION
    - *Nervous Control*
        - *Medullary Rhythmicity Area*
        - *Pneumotaxic Area*
        - *Apneustic Center*
- REGULATION OF RESPIRATORY CENTER ACTIVITY
    - *Cortical Influences*
    - *Inflation Reflex*

- *Chemical Stimuli*
- *Other Influences*
- COMMON DISORDERS
- MEDICAL TERMINOLOGY AND CONDITIONS
- WELLNESS FOCUS: GIVING VIRUSES A COLD RECEPTION

I. CHAPTER SYNOPSIS

This chapter details the structure and function of the organs of respiration. After detailing both gross and microscopic anatomy and some histology, the mechanics and physiology of respiration are considered. Attention is given to pulmonary ventilation, pulmonary air volumes and capacities, the exchange of respiratory gases, external respiration, and internal respiration. Considerable discussion entails the transportation and chemical interaction of oxygen and carbon dioxide. The control of respiration is detailed by discussing the roles of the medulla and pons. Attention is then directed toward the factors which regulate respiratory activity. Those considered are the cerebral cortex, inflation reflex, chemical stimuli, and several other minor factors. The chapter concludes with a list of common disorders, medical terms, and conditions associated with the respiratory system.

II. LEARNING GOALS/STUDENT OBJECTIVES

1. Describe the structure of the nose and how it functions in breathing.
2. Describe the structure of the pharynx and how it functions in breathing.
3. Describe the structure of the larynx (voice box) and explain how it functions in breathing and voice production.
4. Describe the structure and functions of the trachea and bronchi.
5. Describe the structure of the lungs and their role in breathing.
6. Explain how inspiration (breathing in) and expiration (breathing out) take place.
7. Explain how oxygen and carbon dioxide are exchanged between the lungs and blood, and between blood and body cells.
8. Describe how the blood transports oxygen and carbon dioxide.
9. Explain how the nervous system controls breathing and list the factors that can alter the rate of breathing.

III. SAMPLE LECTURE OUTLINE

A. ORGANS

1. The respiratory organs include the NOSE, PHARYNX, LARYNX, TRACHEA, BRONCHI, and LUNGS. They act with the cardiovascular system to supply oxygen and remove carbon dioxide from the blood.

B. Nose

1. The EXTERNAL NOSE is composed of cartilage and skin and is lined with mucous membranes. Openings to the exterior are referred to as the EXTERNAL NARES. The internal portion communicates with the PARANASAL SINUSES and nasopharynx through the INTERNAL NARES.
2. The NASAL CAVITY is divided by a SEPTUM and the anterior portion of the cavity is called the VESTIBULE.
3. The nose is adapted for warming, moistening, and filtering air. It also receives olfactory stimuli and provides large, hollow sinuses as resonating chambers for speech sounds.

C. Pharynx

1. The PHARYNX (throat) is a muscular tube lined with mucous membranes and is divided into three anatomical regions: the nasopharynx, oropharynx, and laryngopharynx.
2. The PHARYNX serves as a passageway for air and food and provides an additional resonating chamber for sound.
3. The LARYNX (voice box) is a passageway that connects the pharynx with the trachea, and contains several cartilages: THYROID CARTILAGE, EPIGLOTTIS, CRICOID CARTILAGE, PAIRED CORNICULATE, ARYTENOID and CUNEIFORM CARTILAGES.
4. The larynx contains VOCAL FOLDS, which produce sound, tightened folds produce high pitches, and relaxed folds produce low pitches.

D. Trachea and Bronchi

1. The TRACHEA (windpipe) extends from the larynx to the PRIMARY BRONCHI. It is composed of smooth muscle and supportive C-rings of cartilage, and is lined with pseudostratified ciliated epithelium.
2. The BRONCHIAL TREE consists of the TRACHEA, PRIMARY BRONCHI, SECONDARY BRONCHI, TERTIARY BRONCHI, BRONCHIOLES, and TERMINAL BRONCHIOLES.

E. Lungs

1. The bronchiole tree functions in transporting air from the trachea to the alveoli of the lungs.
2. The LUNGS are paired organs in the thoracic cavity enclosed by paired PLEURAL MEMBRANES. The PARIETAL PLEURAE lines the thoracic cavity, and the VISCERAL PLEURAE covers each lung.
3. Between the visceral and parietal pleura is a small, potential space, called the PLEURAL CAVITY, which contains a lubricating, serous fluid which functions to reduce friction between the pleurae.
4. The right lung has three lobes separated by FISSURES; the left lung has two lobes separated by a fissure and a depression called the cardiac notch. Each lobe contains its own SECONDARY BRONCHI.
5. The SECONDARY BRONCHI give rise to branches called SEGMENTAL BRONCHI, which supply segments of the lung tissue called BRONCHOPULMONARY SEGMENTS.
6. Each segment consists of many small compartments called LOBULES. Each lobule contains lymphatic tissue, arterioles, venules, terminal bronchioles, respiratory bronchioles, alveolar ducts, alveolar sacs, and alveoli.

7. Gas exchange occurs by diffusion across the ALVEOLAR-CAPILLARY MEMBRANE, which consists of two single layers of SQUAMOUS EPITHELIUM (one alveolar and one capillary) between which lie epithelial and capillary basement membranes, which are often fused to one another.
8. The lungs have a double blood supply. Blood enters the lung via the PULMONARY ARTERIES of pulmonary circulation and BRONCHIOLE ARTERIES of systemic circulation. Blood leaves via the pulmonary veins.

F. RESPIRATION

1. PULMONARY VENTILATION, or breathing, is the first of three basic processes. Pulmonary ventilation includes INSPIRATION (breathing in) and EXPIRATION (breathing out). The other two processes are EXTERNAL RESPIRATION and INTERNAL RESPIRATION.
2. The movement of air into and out of the lungs depends on pressure changes governed by BOYLE'S LAW, which states that the volume of a gas varies inversely with pressure at a constant temperature.
3. INSPIRATION occurs when INTRAPULMONIC PRESSURE (pressure within the alveoli) falls below atmospheric pressure. Contraction of the diaphragm and rib muscles (external intercostals) increases the size of the thorax, thus decreasing the INTRAPLEURAL or INTRATHORACIC PRESSURE (pressure in the thoracic cavity) so that the lungs expand. Expansion of the lungs decreases intrapulmonic pressure so that air moves along the pressure gradient from the atmosphere into the lungs.
4. EXPIRATION occurs when INTRAPULMONIC PRESSURE is higher than atmospheric pressure. Relaxation of the diaphragm and rib muscles results in an elastic recoil of the chest wall and lungs which increases intrathoracic pressure. Lung volume decreases and INTRAPULMONIC PRESSURE increases, so air moves from the lungs to the atmosphere.
5. COMPLIANCE is the ease with which the lungs and thoracic wall can expand.
6. The walls of the respiratory passageways, especially the bronchi and bronchioles, offer some resistance to the normal flow of air into the lungs.
7. SURFACTANT, a phospholipid, is produced in the lungs and functions to reduce surface tension and friction of air passing over the alveoli.

G. PULMONARY AIR VOLUMES AND CAPACITIES

1. Air volumes exchange and rate of ventilation are measured with an instrument called the SPIROMETER.
2. Among the pulmonary air volumes exchanged in ventilation are TIDAL (500 cc), INSPIRATORY RESERVE (3100 cc), EXPIRATORY RESERVE (1200 cc), RESIDUAL (1200 cc), and MINIMAL VOLUME. Only about 350 cc of the tidal volume reaches the alveoli. The remaining 150 cc remains in the airways as anatomic DEAD SPACE.
3. Pulmonary lung capacities, the sum of two or more volumes, include RESPIRATORY CAPACITY (3600 cc), FUNCTIONAL RESIDUAL CAPACITY (2400 cc), VITAL CAPACITY (4800 cc), and TOTAL LUNG CAPACITY (6000 cc).
4. The MINUTE VOLUME of respiration is the total air taken in per minute computed by the product of the tidal volume and the number of respirations per minute.

H. EXCHANGE OF RESPIRATORY GASES

1. The exchange of oxygen and carbon dioxide between the blood and the alveoli is dependent upon several gas laws. These are CHARLES' LAW, DALTON'S LAW, and HENRY'S LAW.
2. CHARLES' LAW states that the volume of a gas is directly proportional to the absolute temperature, assuming that pressure remains constant. As gases enter the warmed lungs, the gases expand, thus increasing lung volume.
3. DALTON'S LAW states that each gas in a mixture of gases exerts its own pressure as if all other gases were not present. This partial pressure of a gas, the pressure exerted by that gas in a mixture of gases, is symbolized *p,* as in pO_2 or pCO_2.
4. HENRY'S LAW states that the quantity of a gas that will dissolve in a liquid is proportional to the partial pressure of the gas and its solubility coefficient when the temperature remains constant.

I. EXTERNAL AND INTERNAL RESPIRATION

1. In EXTERNAL and INTERNAL RESPIRATION, oxygen and carbon dioxide each move from areas of higher partial pressures to areas of lower partial pressures.
2. EXTERNAL RESPIRATION is the exchange of oxygen and carbon dioxide between the alveoli and the pulmonary blood capillaries. It results in the conversion of deoxygenated blood coming from the heart to oxygenated blood returning to the heart.
3. EXTERNAL RESPIRATION is aided by a thin alveolar-capillary (respiratory) membrane, a large alveolar surface area, and a rich blood supply. Its efficiency is dependent on altitude, total surface area for oxygen-carbon dioxide exchange, minute volume, and diffusion distance.
4. INTERNAL RESPIRATION is the exchange of oxygen and carbon dioxide between the blood capillaries and tissue cells. It results in the conversion of oxygenated blood into deoxygenated blood.
5. Approximately 25% of the available oxygen in oxygenated blood actually enters tissue cells when the body is at rest. This increases proportionately as exercise increases.

J. TRANSPORT OF RESPIRATORY GASES

1. In each 100 ml of oxygenated blood, 3% of the oxygen is dissolved in the plasma and 97% is carried in chemical combination with hemoglobin as OXYHEMOGLOBIN.
2. HEMOGLOBIN consists of a protein portion called GLOBIN and a pigment portion called HEME. The heme portion contains four atoms of iron, each capable of combining with a molecule of oxygen.
3. The ASSOCIATION of oxygen and hemoglobin is affected by the PARTIAL PRESSURES of OXYGEN and CARBON DIOXIDE, TEMPERATURE, and 2,3-DIPHOSPHOGLYCERATE (2,3 DPG).
4. The greater the partial pressure of oxygen, the more oxygen will combine with hemoglobin, until all available hemoglobin molecules are saturated. This is generally represented by an oxygen-hemoglobin association curve.
5. In an acid environment, oxygen splits more readily from the hemoglobin. This is referred to as the BOHR EFFECT. Low blood pH results from high partial pressures of carbon dioxide.
6. As temperature increases, so does the amount of oxygen released from the hemoglobin molecule. Cells require more oxygen and active cells liberate more heat and acid, which in turn, increase oxygen-hemoglobin dissociation.

7. 2,3-DIPHOSPHOGLYCERATE is a substance formed in the red blood cells during glycolysis. The greater the level of DPG, the more oxygen is released from the hemoglobin.
8. Fetal hemoglobin has a higher affinity for oxygen than adult hemoglobin and can carry more oxygen to offset low oxygen saturation in maternal blood in the placenta.
9. In each 100 ml of deoxygenated blood, 7% of the carbon dioxide is dissolved in the plasma, 23% combines with the globin portion of the hemoglobin molecule as CARBAMINOHEMOGLOBIN, and 70% is converted to bicarbonate ion.
10. The conversion of carbon dioxide to BICARBONATE ions and the related chloride shift maintains the ionic balance between red blood cells and plasma.
11. Carbon dioxide in the blood causes oxygen to split from hemoglobin. Similarly, the binding of oxygen to hemoglobin causes a release of carbon dioxide from the blood. In the presence of oxygen, less carbon dioxide bonds in the blood, a reaction called the HALDANE EFFECT.

K. CONTROL OF RESPIRATION

1. The area of the brain from which nerve impulses are sent to the respiratory muscles is located bilaterally in the RETICULAR FORMATION of the brain stem. The respiratory center consists of a MEDULLARY RHYTHMICITY CENTER (inspiratory and expiratory areas), PNEUMOTAXIC AREA, and APNEUSTIC AREA.
2. The MEDULLARY RHYTHMICITY AREA controls the basic rhythm of respiration.
3. The INSPIRATORY AREA has an intrinsic excitability that sets the basic rhythm of respiration.
4. Neurons of the EXPIRATORY AREA remain inactive during most quiet respiration but are probably excited during high levels of ventilation to cause contraction of the muscles used to force expiration.
5. The PNEUMOTAXIC and APNEUSTIC AREAS are located in the PONS, and coordinate the transition between inspiration and expiration.
6. The PNEUMOTAXIC AREA sends impulses that limit inspiration and facilitate expiration, preventing overexpansion of the lungs.
7. The APNEUSTIC AREA sends impulses to the inspiratory area that activates it and prolongs inspiration, thus inhibiting expiration.

L. REGULATION OF RESPIRATORY CENTER ACTIVITY

1. The rhythm of respiration may be modified by cortical influences, the inflation reflex, chemical stimuli as oxygen and carbon dioxide, temperature, pain, pressure, and irritation of the respiratory mucosa.
2. The modifying factors of respiration are cortical influences, the inflation reflex, chemical stimuli, O_2 and CO_2 levels, blood pressure, temperature, pain, irritation to the respiratory mucosa, and exercise.

IV. TEACHING TIPS AND SUGGESTIONS

A. HELPFUL HINTS

1. Have students demonstrate the muscular contractions of normal and forced expirations and inspirations on themselves. Point out the magnitude of chest and abdominal movement with

forced inspiration and the more noticeable change in the chest and abdominal size in forced expiration. Have the student explain these changes in size on the basis of the muscles involved and their actions.

2. Use a spirometer to compare the lung volumes of smokers vs. non-smokers, large individuals vs. small individuals, and older vs. younger students.
3. In discussing neuroregulation, emphasize that the regulation of respiratory center activity is influenced to a much greater extent by hydrogen ions associated with high partial pressure of carbon dioxide than by high partial pressure of oxygen. Chemoreceptors in the carotid and aortic bodies outside the central nervous system are finely tuned to hydrogen levels, sending signals to the inspiratory center to increase ventilation rates when the partial pressure of carbon dioxide level is above 40 TORR.

B. ESSAYS

1. Plot the course taken by molecules of oxygen to the heart and the reverse course of carbon dioxide from the heart to the atmosphere. Diagram and explain the movements of these gases.
2. A 5-year old child threatens to hold her breath if the parent does not give her an ice cream cone. Would you advise the parent to let the child hold her breath or to give in to the demand? Your advice should be based on any possible physiological damage to the child. Explain what would happen if the child chose to hold her breath.

C. TOPICS FOR DISCUSSION

1. Discuss the basic steps involved in inspiration and expiration. Include all pressures involved.
2. Discuss how the control of respiration demonstrates the principle of homeostasis.

V. AUDIOVISUAL MATERIALS

A. OVERHEAD TRANSPARENCIES

1. PAP Transparency Set (TRS. 23.1a&b, 23.4, 23.7a, 23.9a-d, 23.10, 23.11a&b, 23.12a&b, 23.13a-c, 23.14a-c, 23.15a&b, 23.16-23.23, 23.25a&b, 23.26-23.28 & 23.30).
2. Respiratory System (HSC).
3. Respiratory System: Structure (GAF).
4. Respiratory System: Unit 7 (10 Transparencies; RJB).

B. VIDEOCASSETTES

1. Breath of Life (26 min.; FHS).
2. The Physiology of Exercise (15 min.; FHS).
3. Smoking: Hazardous to Your Health (29 min.; 1983; KSU).
4. The Coach's Final Lesson (23 min.; 1985; ALA).
5. Asthma (19 min.; C; Sd; 1990; FHS).
6. Cystic Fibrosis (26 min.; C; Sd; 1991; FHS).
7. Smoking: Kicking the Habit (29 min.; C; Sd; 1983; KSU).
8. The Coach's Final Lesson (23 min.; C; Sd; 1984; ALA).
9. The Respiratory System (26 min.; C; Sd; 1989; GA).

10. Up In Smoke: How Smoking Affects Your Health (38 min.; C; Sd; 1990; GA).

C. Films: 16 mm

1. The Human Body: Respiratory System (13 min.; 1980; COR/KSU).
2. I am Joe's Lung (25 min.; 1975; PYR/KSU).
3. Respiration in Man (25 min.; 1985; McG/EBEC/KSU).
4. The Lung and the Respiratory System (16 min.; 1975; EBEC/KSU).
5. New Breath of Life (12 min.; 1974; CRM/KSU).
6. Mechanisms of Breathing (11 min.; EBF).
7. Respiration (27 min.; 1961; McG/KSU).
8. CPR Quiz: Basic Life Support (21 min.; TLV).
9. CPR Trainer (21 min.; TLV).
10. CPR for Citizens (25 min.; 1978; PYR/KSU).
11. CPR: To Save a Life (14 min.; 1977; EBEC/KSU).

D. Transparencies: 35 mm (2x2)

1. PAP Slide Set (Slides 106-114).
2. AHA Slide Set.
3. Visual Approach to Histology: Respiratory System (11 Slides; FAD).
4. Biology: Respiratory System, Breathing (EBF).
5. Histology of the Respiratory System (EIL).
6. Respiratory System and Its Functions (20 Slides; EIL).
7. The Respiratory System Slide Set (210 Slides; CARO).

E. Computer Software

1. Body Language: Respiratory System (Apple II Series; ESP).
2. Circulation and Respiration (Apple; IBM PC; MAC; PLP).
3. Graphic Human Anatomy and Physiology Tutor: Respiratory System (IBM; PLP).
4. Tobacco: To Smoke or Not to Smoke (Apple; PLP).
5. Understanding Human Physiology: Exercise Physiology (IBM; PLP).

CHAPTER 19

THE DIGESTIVE SYSTEM

CHAPTER AT A GLANCE

- DIGESTIVE PROCESSES
- ORGANIZATION
    - *General Histology*
        - *Mucosa*
        - *Submucosa*
        - *Muscularis*
        - *Serosa and Peritoneum*
- MOUTH (ORAL) CAVITY
    - *Tongue*
    - *Salivary Glands*
        - *Composition of Saliva*
        - *Secretion of Saliva*
    - *Teeth*
- DIGESTION IN THE MOUTH
    - *Mechanical*
    - *Chemical*
- PHARYNX
- ESOPHAGUS
    - *Swallowing*
- STOMACH
    - *Anatomy*
    - *Digestion in the Stomach*
        - *Mechanical*
        - *Chemical*
    - *Regulation of Gastric Secretion*
        - *Stimulation*
        - *Inhibition*
    - *Regulation of Gastric Emptying*
    - *Absorption*
- PANCREAS
    - *Anatomy*
    - *Pancreatic Juice*
    - *Regulation of Pancreatic Secretions*
- LIVER
    - *Anatomy*
    - *Blood Supply*
    - *Bile*
    - *Regulation of Bile Secretion*
    - *Functions of the Liver*

- GALLBLADDER
  - *Functions*
  - *Emptying of the Gallbladder*
- SMALL INTESTINE
  - *Anatomy*
  - *Intestinal Juice*
  - *Digestion in the Small Intestine*
    - *Mechanical*
    - *Chemical*
  - *Regulation of Intestinal Secretion and Mobility*
  - *Absorption*
    - *Carbohydrate Absorption*
    - *Protein Absorption*
    - *Lipid Absorption*
    - *Water Absorption*
    - *Vitamin Absorption*
- LARGE INTESTINE
  - *Anatomy*
  - *Digestion in the Large Intestine*
    - *Mechanical*
    - *Chemical*
  - *Absorption and Feces Formation*
  - *Defecation*
- COMMON DISORDERS
- MEDICAL TERMINOLOGY AND CONDITIONS
- WELLNESS FOCUS: ALCOHOL USE OR ABUSE?

I. CHAPTER SYNOPSIS

In this chapter, the student is introduced to the major structures and functions of the digestive system and its organization into the gastrointestinal tract and accessory structures. A detail of the general histology of the layers of the digestive tract is presented. The entire digestive process is considered by region, and the structure, function, physical processes, and chemical processes are discussed together. Emphasis is placed on the neural and hormonal mechanisms regulating the control of digestive secretions. The chapter concludes with a list of common disorders, and medical terms, and conditions associated with the digestive system.

II. LEARNING GOALS/STUDENT OBJECTIVES

1. Classify the organs of the digestive system and describe the histology of the gastrointestinal tract.
2. Describe the structure of the mouth and explain its role in digestion.
3. Describe the stages involved in swallowing.
4. Explain the structure and functions of the stomach in digestion.
5. Describe the structure of the pancreas, liver, and gallbladder, and explain their functions in digestion.

6. Explain how the small intestine is adapted for digestion and absorption.
7. Describe the structure of the large intestine and explain its function in digestion, feces formation, and defecation.

III. SAMPLE LECTURE OUTLINE

A. DIGESTIVE PROCESSES

1. Food is prepared for use by cells through five basic activities: ingestion, movement of food along the gastrointestinal tract, mechanical and chemical digestion, absorption, and defecation.
2. CHEMICAL DIGESTION is a series of catabolic reactions that break down large carbohydrate, lipid, and protein molecules into smaller molecules that can be absorbed and used by body cells.
3. MECHANICAL DIGESTION consists of various movements that increase surface area to aid chemical digestion.
4. ABSORPTION is the passage of the end products of digestion from the gastrointestinal tract into the blood or lymph for distribution to cells.
5. DEFECATION is the emptying of the rectum, eliminating indigestible materials from the gastrointestinal tract.

B. ORGANIZATION

1. The organs of digestion are divided into two main groups: those composing the GASTROINTESTINAL (GI) TRACT, or alimentary canal, and those composing the ACCESSORY STRUCTURES.
2. The GI TRACT is a continuous tube running from the mouth to the anus through the ventral cavity of the body. The organs involved are the MOUTH, PHARYNX, ESOPHAGUS, STOMACH, SMALL INTESTINE, and LARGE INTESTINE.
3. The GI tract contains the food from the time it is ingested until it is eliminated from the body.
4. The ACCESSORY STRUCTURES include the TEETH, TONGUE, SALIVARY GLANDS, LIVER, GALLBLADDER, and PANCREAS.
5. The basic arrangement of LAYERS in the GI tract from the inside outward is MUCOSA, SUBMUCOSA, MUSCULARIS, and SEROUS OR ADVENTITIA. The outermost layer is also called the VISCERAL PERITONEUM.
6. The PERITONEUM is the largest of the serous membranes. The PARIETAL PERITONEUM lines the wall of the abdominal cavity, and the VISCERAL PERITONEUM covers some of the organs and constitutes their serosa.
7. The potential space between the parietal and visceral peritoneum is called the PERITONEAL CAVITY and contains serous fluid.
8. Some organs, such as the pancreas and kidneys, lie on the posterior abdominal wall behind the peritoneum and are referred to as a RETROPERITONEAL ORGANS.
9. The peritoneum contains large folds that weave between the viscera, functioning to support the organs and blood vessels, lymphatic vessels, and nerves to the abdominal organs. These extensions of the peritoneum include the MESENTERY, MESOCOLON, FALCIFORM LIGAMENT, LESSER OMENTUM, and GREATER OMENTUM.

C. Mouth (Oral Cavity)

1. The MOUTH, also called the oral or buccal cavity, is formed by the cheeks, hard and soft palates, lips, and tongue.
2. The OPENING of the oral cavity is bounded externally by the cheeks and lips, and internally by the gums (gingiva) and teeth.
3. The oral cavity proper extends from the opening (vestibule) to the opening between the oral cavity and pharynx (fauces).
4. The TONGUE, together with its associated muscles, forms the flow of the oral cavity and is composed of skeletal muscle covered with mucous membrane.
5. Extrinsic and intrinsic muscles permit the tongue to be moved and participate in food manipulation for chewing (MASTICATION) and swallowing (DEGLUTITION).
6. The upper surface and the sides of the tongue are covered with papillae, some of which contain the taste buds.

D. Salivary Glands

1. The SALIVARY GLANDS lie outside the oral cavity and secrete the major portion of saliva through ducts which enter into the mouth. The remainder of the saliva comes from buccal glands in the mucous membrane that line the mouth.
2. There are THREE PAIRS OF SALIVARY GLANDS: PAROTID, SUBMANDIBULAR (SUBMAXILLARY), and SUBLINGUAL GLANDS.
3. SALIVA lubricates and dissolves food and starts the chemical digestion of carbohydrates through the action of salivary amylase. It also functions to keep the mucous membranes of the mouth and throat moist.
4. Chemically, saliva is 99.5% water. The remaining 0.5% is composed of solutes such as salts, dissolved gases, various organic substances, and enzymes. The secretion of saliva is under the control of the nervous system.

E. Teeth

1. The TEETH project into the mouth and are adapted for mechanical digestion. The typical tooth consists of three principal parts: crown, root, and neck.
2. Teeth are COMPOSED primarily of DENTIN. The crowns of the teeth are covered by enamel, the hardest substance in the body.
3. The dentin of the root is covered by cementum, a bone-like substance which attaches the root of the tooth to the periodontal ligament.
4. The dentin encloses the pulp cavity in the crown and the root canals in the roots.
5. The DECIDUOUS or PRIMARY TEETH are replaced by the permanent or secondary teeth.
6. There are four different types of teeth based on shape: INCISORS, used to cut food; CANINES, used to tear food; PREMOLARS or BICUSPIDS, used for crushing and grinding food, and MOLARS, used for crushing and grinding food.

F. Physiology of Digestion in the Mouth

1. Through mastication, food is mixed with saliva and shaped into a bolus that can easily be swallowed.

2. The enzyme SALIVARY AMYLASE converts polysaccharides (starches) to a disaccharides (maltose). This is the only chemical digestion that occurs in the mouth.
3. DEGLUTITION, or the act of swallowing, is a mechanism that moves the bolus from the mouth to the stomach. It is facilitated by saliva and mucus and involves the mouth, pharynx, and esophagus.
4. Deglutition consists of a voluntary stage, an involuntary pharyngeal stage, and involuntary esophageal stage.

G. ESOPHAGUS

1. The ESOPHAGUS is a collapsible, muscular tube that lies dorsal to the trachea and connects the pharynx to the stomach.
2. The wall of the esophagus contains a mucosa, submucosa, and muscularis layer. The outermost layer is called adventitia rather than serosa due to structural differences.
3. The esophagus passes the bolus to the stomach by peristalsis, and contains the upper and lower esophageal sphincters.

H. STOMACH ANATOMY

1. The STOMACH is a J-shaped enlargement of the GI tract which begins at the CARDIAC VALVE at the bottom of the esophagus and ends at the PYLORIC SPHINCTER.

2. The gross anatomical subdivisions of the stomach are the CARDIA, FUNDUS, BODY, and PYLORUS.
3. The stomach contains mucosal folds called RUGAE and numerous glands that produce mucus, hydrochloric acid, proteinase, intrinsic factor, and stomach gastrin. The stomach also contains a three-layered muscularis for efficient mechanical movement.
4. The GASTRIC GLANDS consists of four types of secreting cells: PEPTIC CELLS, PARIETAL CELLS, MUCOUS CELLS, and ENTEROENDOCRINE CELLS.
5. The PEPTIC CELLS release a precursor enzyme, PEPSINOGEN. The PARIETAL CELLS secrete HCl, which aids in the conversion of pepsinogen to active pepsin, and intrinsic factor, involved in the absorption of Vitamin B-12. The MUCUS CELLS secrete mucus, and the ENTEROENDOCRINE cells secrete stomach GASTRIN, a hormone which allows for the secretion of HCl and pepsinogen, contracts the lower esophageal sphincter, and increases peristalsis.

I. STOMACH PHYSIOLOGY

1. MECHANICAL DIGESTION in the stomach consists of PERISTALTIC MOVEMENTS called mixing waves.
2. CHEMICAL DIGESTION consists of the conversion of protein into peptides by the action of the enzyme pepsin. Pepsin is converted from its inactive form, pepsinogen, through the secretion of HCl.
3. The resulting material from the chemical and mechanical gastric digestion is called CHYME.
4. GASTRIC SECRETION is REGULATED by hormonal and neural mechanisms, and its stimulation occurs in three phases: CEPHALIC, GASTRIC, and INTESTINAL.
5. When partially digested food enters the small intestine, it triggers the ENTEROGASTRIC REFLEX and the secretion of several hormones: SECRETIN, CHOLECYSTIKININ (CCK), and GASTRIC-

INHIBITING PEPTIDE (GIP). These are released by the intestinal mucosa and inhibit gastric secretion.
6. Gastric emptying is stimulated by two factors: nerve impulses in response to distention of the stomach, and release of gastrin by the stomach in response to the presence of certain types of food.
7. Most food leaves the stomach two to six hours after ingestion; carbohydrates leave the fastest, followed by proteins, and then fats.
8. Gastric emptying is INHIBITED by the ENTEROGASTRIC REFLEX and the GASTRIC HORMONES.
9. The stomach is permeable only to water, electrolytes, certain drugs, aspirin, and alcohol.

J. PANCREAS

1. The PANCREAS is an accessory organ which functions as both an endocrine organ and an exocrine organ. The exocrine, or digestive functions of the pancreas, involve the secretion of pancreatic juice into the small intestine via ducts.
2. The PANCREATIC JUICE contains enzymes that digest starch (pancreatic amylase), proteins (trypsin, chymotrypsin, and carboxypeptidase), fats (pancreatic lipase), and nucleotides (nucleases).
3. Under the influence of SECRETIN, the pancreas secretes BICARBONATE IONS, which convert the stomach acid contents to a slightly alkaline pH (7.1-8.2). This also inhibits stomach pepsin activity and promotes the activity of the pancreatic enzymes.
4. Pancreatic secretions are regulated by nervous and hormonal mechanisms.

K. LIVER AND GALLBLADDER

1. The LIVER is divided into right and left lobes. The larger, right lobe is further subdivided into the caudate and quadrate lobes.
2. The LOBES of the liver are made up of lobules that contain liver cells called HEPATOCYTES, SINUSOIDS, and KUPFFER'S CELLS, and a CENTRAL VEIN.
3. The liver receives a dual blood supply from the HEPATIC ARTERY and the HEPATIC PORTAL VEIN. All blood eventually leaves the liver via the hepatic vein.
4. The HEPATOCYTES produce bile that is transported by a duct system to the gallbladder for concentration and temporary storage.
5. BILE is partially an excretory product and digestive secretion, and functions in the emulsification of fats.
6. The rate of bile secretion is regulated by nervous and hormonal mechanisms as well as by the volume of hepatic flow and the concentration of bile salts in the blood.
7. The liver also functions in carbohydrate, fat, and protein metabolism, removal of drugs and hormones, excretion of bile, synthesis of bile salts, storage of vitamins and minerals, phagocytosis, and activation of Vitamin D.
8. The GALLBLADDER is a sac located in a fossa of the visceral surface of the liver. It functions to store and concentrate the bile produced by the liver.
9. BILE is ejected, under chemical stimulation from CCK, into the common bile duct which leads to the duodenum.

L. SMALL INTESTINE ANATOMY

1. The SMALL INTESTINE extends from the pyloric sphincter to the ileocecal sphincter and is divided into the DUODENUM, JEJUNUM, and ILEUM.
2. Highly adapted for digestion and absorption, the small intestine contains glands that produce enzymes and mucus. The MICROVILLI, VILLI, and CIRCULAR FOLDS on its wall provide a large surface area for digestion and absorption.
3. INTESTINAL JUICE provides a means for the absorption of substances from chyme as they come in contact with the villi. Some of the intestinal enzymes break down foods inside epithelial cells of the mucosa on the surfaces of their microvilli. Some digestion occurs in the lumen of the small intestine.

M. DIGESTION IN THE SMALL INTESTINE

1. MECHANICAL DIGESTION in the small intestine involves SEGMENTATION and PERISTALSIS.
2. Intestinal enzymes include MALTASE, SUCRASE, LACTASE, DIPEPTIDASES, and NUCLEASES.
3. LIPIDS are first emulsified by bile salts and then hydrolyzed to fatty acids and monoglycerides by pancreatic lipase.
4. Local reflexes in response to the presence of chyme regulate small intestinal secretions. The hormones secretin and cholecystikinin also assume a role in this mechanism.
5. MONOSACCHARIDES, AMINO ACIDS, and SHORT-CHAINED FATTY ACIDS pass into the blood capillaries.
6. The small intestine also ABSORBS WATER, ELECTROLYTES, and VITAMINS.
7. LONG-CHAINED FATTY ACIDS and MONOGLYCERIDES are dissolved in the center of aggregated bile salts called MICELLES.
8. When the MICELLES reach the epithelial cells of the intestinal villi, the fatty acids and monoglycerides diffuse into the cells, leaving the micelles behind. The monoglycerides are digested into GLYCEROL and FATTY ACIDS, and then recombine to form TRIGLYCERIDES.
9. The TRIGLYCERIDES combine with PHOSPHOLIPIDS and CHOLESTEROL to form protein-coated masses called CHYLOMICRONS. The chylomicrons are taken up by the LACTEALS of the villus, enter the lymphatic and cardiovascular systems, and finally reach the liver or fatty tissue.
10. The plasma lipids such as FATTY ACIDS, TRIGLYCERIDES, and CHOLESTEROL are insoluble in water. In order to be transported in blood and utilized by cells, the lipid must combine with protein transporters called APOPROTEINS to make them soluble.
11. This combination of lipid and protein is referred to as a LIPOPROTEIN.
12. In addition to CHYLOMICRONS, there are several other types of lipoproteins, known as HIGH-DENSITY LIPOPROTEINS (HDLs), LOW-DENSITY LIPOPROTEINS (LDLs), and VERY LOW-DENSITY LIPOPROTEINS (VLDLs).
13. The VLDLs function to transport triglycerides synthesized in the liver to adipose and muscle tissue where the VLDLs are catabolized and triglycerides are deposited.
14. LDLs account for 60-70% of the total serum cholesterol. The function of LDLs is steroid synthesis, bile acid production, and the manufacturing of cell membranes.
15. HDLs are rich in phospholipids and cholesterol, and transport 20-30% of the total serum cholesterol. Their function is to transport cholesterol from peripheral tissues to the liver where it is catabolized to become a component of bile. High HDL levels are associated with a decreased risk of cardiovascular disease since they remove cholesterol from arterial walls.

N. Large Intestine Anatomy

1. The LARGE INTESTINE extends from the ILEOCECAL SPHINCTER to the ANUS.
2. Its subdivisions include the CECUM, ASCENDING COLON, TRANSVERSE COLON, DESCENDING COLON, SIGMOID COLON, RECTUM, and ANAL CANAL.
3. The mucosa of the large intestine has no villi or permanent circular folds. It does contain simple columnar epithelium with numerous goblet cells, and the muscularis contains specialized portions of longitudinal muscles called TAENIAE COLI.

O. Digestion in the Large Intestine

1. MECHANICAL MOVEMENTS of the large intestine include HAUSTRAL CHURNING, PERISTALSIS, and MASS PERISTALSIS.
2. The last stages of chemical digestion occur in the large intestine through bacterial action. Some mechanical digestion occurs and Vitamin K is synthesized and absorbed.
3. The large intestine ABSORBS WATER, ELECTROLYTES, and some VITAMINS.
4. The FECES consist of water, inorganic salts, sloughed-off epithelial cells from the mucosa of the GI tract, bacteria, products of bacterial decomposition, and undigested food materials.
5. The ELIMINATION of FECES from the rectum is called DEFECATION.
6. DEFECATION is a reflex action aided by voluntary contractions of the diaphragm and abdominal muscles.

P. Common Disorders

1. DENTAL CARIES are started by acid-producing bacteria that reside in dental plaque and demineralize tooth enamel with acid.
2. PERIODONTAL DISEASES are characterized by inflammation and degeneration of the gingivae, alveolar bone, periodontal membrane, and cementum.
3. PEPTIC ULCERS are lesions that develop in the mucus membranes of the GI tract in areas exposed to gastric juice.
4. APPENDICITIS is an inflammation of the veriform appendix resulting from obstruction of the lumen of the appendix by impaction, stenosis, or kinking of the organ.
5. DIVERTICULI are outpouchings of the colonic wall where the muscularis has weakened.
6. ANOREXIA nervosa is a chronic disorder characterized by self-induced weight loss, due to negative body image or other perceptual disturbances, and physiologic changes that result from malnutrition.
7. BULEMIA, or binge-purge syndrome, involves uncontrollable overeating followed by self-induced vomiting, excessive exercise, and abuse of laxatives.

MAJOR HORMONES THAT CONTROL DIGESTION

HORMONE	WHERE PRODUCED	STIMULUS FOR SECRETION
GASTRIN	MUCOSA OF PYLORIC ANTRUM; SMALL QUANTITY FROM INTESTINAL MUCOSA	DISTENTION OF THE STOMACH; PARTIALLY DIGESTED PROTEINS AND CAFFEINE IN STOMACH; AND A HIGH pH OF STOMACH CHYME
GASTRIC INHIBITORY PEPTIDE	INTESTINAL MUCOSA	FATTY ACIDS AND GLUCOSE THAT ENTER THE SMALL INTESTINE
SECRETIN	INTESTINAL MUCOSA	ACIDIC CHYME ENTERING THE SMALL INTESTINE
CHOLECYSTIKININ	INTESTINAL MUCOSA; BRAIN	PARTIALLY DIGESTED PROTEINS AND TRIGLYCERRIDES THAT ENTER THE SMALL INTESTINE

SUMMARY OF DIGESTIVE ENZYMES

ENZYME	SOURCE	SUBSTRATE
SALIVARY AMYLASE	SALIVARY GLANDS	STARCHES
LINGUAL LIPASE	GLANDS IN THE TONGUE	TRIGLYCERIDES AND OTHER LIPIDS
PEPSIN	STOMACH CHIEF CELLS	PROTEINS
PANCREATIC AMYLASE	PANCREATIC ACINAR CELLS	STARCHES
TRYPSIN	PANCREATIC ACINAR CELLS	PROTEINS
CHYMOTRYPSIN	PANCREATIC ACINAR CELLS	PROTEINS
CARBOXYPEPTIDASE	PANCREATIC ACINAR CELLS	TERMINAL AMINO ACID AT CARBOXYL END OF PEPTIDES
PANCREATIC LIPASE	PANCREATIC ACINAR CELLS	EMULSIFIED TRIGLYCERIDES
RIBONUCLEASE	PANCREATIC ACINAR CELLS	RIBONUCLEIC ACID
DEOXYRIBONUCLEASE	PANCREATIC ACINAR CELLS	DEOXYRIBONUCLEIC ACID
MALTASE	SMALL INTESTINE	MALTOSE
SUCRASE	SMALL INTESTINE	SUCROSE
LACTASE	SMALL INTESTINE	LACTOSE
AMINOPEPTIDASE	SMALL INTESTINE	TERMINAL AMINO ACIDS; ATAMINE END OF PEPTIDE
DIPEPTIDASE	SMALL INTESTINE	DIPEPTIDES
NUCELOSIDASES	SMALL INTESTINE	NUCLEOTIDES
PHOSPHATASES	SMALL INTESTINE	NUCELOTIDES

IV. TEACHING TIPS AND SUGGESTIONS

A HELPFUL HINTS

1. When discussing dentition, explain why dental hypersensitivity is normal.
2. It is useful to detail why liver function tests are often done with a routine blood workup and why the information it reveals is important not only to the functions of the blood but also of the liver.
3. If possible, supplement this lecture in the laboratory with the dissection of an organism to detail the major organs, accessory structures, and sphincters.

B. ESSAYS

1. You have just eaten a sausage and egg sandwich for breakfast. Explain the chemical changes that occur as it passes through the mouth, stomach, small intestine, colon, rectum, and anus. Include all enzymes and secretions that are involved.
2. Explain how gastric emptying is stimulated and inhibited.
3. Describe and compare the composition of pancreatic and intestinal juice.

C. TOPIC FOR DISCUSSION

1. Discuss why someone can survive without a colon. Detail how the rest of the digestive tract compensates for lower digestive functioning.

V. AUDIOVISUAL MATERIALS

A. OVERHEAD TRANSPARENCIES

1. PAP Transparency Set (Trs. 24.1-24.1, 24.11a, 24.12a&b, 24.13, 24.14, 24.16-24.18, 24.20a&b, 24.21, 24.22a&b, 24.25,a&b, 24,27a&b & 24.28a&b).
2. Digestive System: Unit 8 (11 Transparencies; RJB).
3. Digestive System (HSC).
4. Structure of the Tooth (TSED).
5. The Table of Foods (CARO).
6. The Complete Digestive System (GAF).

B. VIDEOCASSETTES

1. Eating to Live (26 min.; FHS).
2. Breakdown (26 min.; FHS).
3. Dangerous Dieting (HRM).
4. Wasting Away: Understanding Anorexia and Bulimia (40 min.; CFH).
5. The Enigma of Anorexia Nervosa: Parts I-II (CARLE/KSU).
6. Alcohol Addiction (28 min.; C; Sd; 1989; FHS).
7. Animal Nutrition (29 min.; C; Sd; 1990; IM).
8. Dangerous Dieting (HRM).
9. Digestion (29 min.; C; Sd; 1978; IM).

10. Eating to Live (26 min.; C; Sd; 1990; FHS).
11. The Enigma of Anorexia Nervosa (40 min.; C; Sd; 1985; CARLE/KSU).
12. The Stomach and its Disorders (10 min.; C; Sd; 1989; PLP).
13. Things That Go Bump in Your GI Tract (28 min.; C; Sd; 1985; KSU).
14. Wasting Away (40 min.; C; Sd; 1991; GA).

C. FILMS: 16 MM

1. The Human Body: Digestive System (16 min.; 1980; COR/KSU).
2. The Digestive System (17 min.; EBEC).
3. I Am Joe's Stomach (25 min.; 1975; PYR/KSU).
4. I Am Joe's Liver (30 min.; 1984; PYR/KSU).
5. Nutrition: Fueling the Human Machine (18 min.; 1978; BFA).
6. Intravenous Hyper-Alimentation Techniques (21 min.; ABB/KSU).
7. Swallowing (H&B).
8. The Digestive Tract (EFL).
9. The Stomach in Action (H&B).

D. TRANSPARENCIES: 35 MM (2x2)

1. PAP Slide Set (Slides 115-119).
2. AHA Slide set.
3. Visual Approach to Histology: Digestive System (48 Slides; FAD).
4. Digestive System (Slides 47-82, McG).
5. Digestive System and Its Function (20 Slides; EIL).
6. Endoscopy Medical Series (115 Slides; CARO).

E. COMPUTER SOFTWARE

1. Digestive System Probe (Apple; SC-175037; PLP).
2. Teeth Probe (Apple; SC-175036; PLP).
3. Digestive System (Apple; AC-182015; IBM, SC-182016; PLP).
4. Digestion (Apple II; 39-8852; TRS-80; 3908853; CARO).
5. The Digestion Simulator (Apple II; 40-1128; CARO).
6. Nutrition (Apple II; 40-1150; CARO).
7. Body Language: Digestive System (Apple II Series; ESP).
8. Enzyme Investigation (Apple II; CARO).
9. Body Language-Digestive System (Apple; IBM PC; MAC; PLP).
10. Graphic Human Anatomy and Physiology Tutor: Digestive System (IBM; PLP).

CHAPTER 20

METABOLISM

CHAPTER AT A GLANCE

- Nutrients
- Regulation of Food Intake
- Guidelines for Healthy Eating
- Metabolism
  - *Anabolism*
  - *Catabolism*
  - *Metabolism of Enzymes*
  - *Oxidation-Reduction Reactions*
- Carbohydrate Metabolism
  - *Fate of Carbohydrates*
  - *Glucose Catabolism*
    - *Glycolysis*
    - *Krebs' Cycle*
    - *Electron Transport Chain*
  - *Glucose Anabolism*
    - *Glucose Storage and Release*
    - *Formation of Glucose from Proteins and Fats: Gluconeogenesis*
- Lipid Metabolism
  - *Fate of Lipids*
  - *Triglyceride Storage*
  - *Lipid Catabolism: Lipolysis*
    - *Glycerol*
    - *Fatty Acids*
  - *Lipid Anabolism*
- Protein Metabolism
  - *Fate of Proteins*
  - *Protein Catabolism*
  - *Protein Anabolism*
- Regulation of Metabolism
- Minerals
- Vitamins
- Metabolism and Body Heat
  - *Measuring Heat*
  - *Production of Body Heat*
  - *Basal Metabolic Rate (BMR)*
  - *Loss of Body Heat*
    - *Radiation*
    - *Conduction*
    - *Convection*
    - *Evaporation*

- HOMEOSTASIS OF BODY TEMPERATURE REGULATION
    - *Hypothalamic Thermostat*
    - *Mechanisms of Heat Production*
    - *Mechanisms of Heat Loss*
    - *Fever*
- COMMON DISORDERS
- WELLNESS FOCUS: THE METABOLIC REALITIES OF WEIGHT CONTROL

I. CHAPTER SYNOPSIS

Students are introduced to metabolism and nutrition. There is a discussion of the meaning of metabolism, followed by an analysis of enzymes. The physiology of energy production, carbohydrate metabolism, lipid metabolism, and protein metabolism are discussed. The regulation of metabolism is outlined including a discussion of body heat production and basal metabolic rate (BMR). Minerals and vitamins are described according to storage areas, source, and importance in metabolism. The chapter continues with the relationship of food to body heat, mechanisms of heat gain and loss, the regulation of body temperature, and some body temperature abnormalities. The chapter concludes with a list of common disorders and a list of medical terms and conditions associated with metabolism.

II. LEARNING GOALS/STUDENT OBJECTIVES

1. Define metabolism and describe its importance to homeostasis.
2. Explain how the body uses carbohydrates.
3. Explain how the body uses lipids.
4. Explain how the body uses proteins.
5. Explain how the body uses minerals.
6. Explain how the body uses vitamins.
7. Explain how metabolism and body heat are related.

III. SAMPLE LECTURE OUTLINE

A. NUTRIENTS

1. Nutrients are chemical substances that provide energy, act as building blocks in forming new body components, or assist in the functioning of various body processes.
2. There are six major classes of nutrients: CARBOHYDRATES, LIPIDS, PROTEINS, MINERALS, VITAMINS, and WATER.

B. Regulation of Food Intake

1. Two centers in the HYPOTHALAMUS related to the regulation of food intake are the feeding (hunger) center and satiety center. The feeding center is constantly activated but may be inhibited by the satiety center.
2. The feeding and satiety centers are affected by many stimuli, such as glucose, amino acids, lipids, body temperature, distension, and cholecystikinin.

C. Guidelines for Healthy Eating

1. The following is the recommended distribution of calories:
   - 50-60% from carbohydrates, with less than 15% from simple sugars.
   - less than 30% from fats with no more than 10% saturated fats.
   - approximately 12-15% from proteins.
2. The guidelines for healthy eating are:
   - Eat a variety of foods.
   - Maintain a healthy weight.
   - Eat food low in fat, saturated fat, and cholesterol.
   - Eat plenty of vegetables, fruits, and grain products.
   - Use sugars only in moderation.
   - Use salt and sodium only in moderation.
   - Use alcoholic beverages only in moderation.

D. Metabolism

1. METABOLISM refers to all the chemical processes of the body and has two phases: CATABOLISM and ANABOLISM.
2. ANABOLISM consists of a series of synthesis reactions whereby small molecules are built up into larger ones that form the body's structural and functional components. Anabolic reactions require the use of energy.
3. CATABOLISM is the term used for the decomposition of larger molecules into smaller molecules and generally liberates energy.
4. The coupling of energy-requiring and energy-releasing reactions is achieved through ATP.

E. Metabolism and Enzymes

1. Metabolic reactions are catalyzed by enzymes, which are proteins that are very efficient, specific for their substrates, and subject to cellular control.
2. The names of enzymes typically end in the suffix -ASE.
3. A complete enzyme, consisting of an APOENZYME and a COFACTOR, is known as a HOLOENZYME.
4. The essential function of an enzyme is to catalyze chemical reactions. Its action, although not completely understood, is believed to follow a sequence of four steps:
   - The surface of the substrate makes contact with a specific region on the A temporary intermediate compound called an enzyme-substrate complex forms.
   - The substrate molecule is transformed by rearrangement of the existing atoms, breakdown of the substrate molecule, or combination of several molecules.
   - The transformed substrate molecules, now called the products of the reaction, move away from the enzyme molecule, allowing the enzyme to attach to another molecule.

F. OXIDATION-REDUCTION REACTIONS

1. OXIDATION is the removal of electrons of hydrogen ions from a molecule and results in a decrease in the energy content of the molecule. REDUCTION is the addition of electrons of hydrogen ions to a molecule and results in an increase in the energy content of a molecule.
2. Within a cell, oxidation and reduction reactions are always coupled. This coupling of reactions is called OXIDATION-REDUCTION REACTIONS.
3. ATP is generated by SUBSTRATE-LEVEL PHOSPHORYLATION, OXIDATIVE PHOSPHORYLATION, and PHOTOPHOSPHORYLATION.
4. Substances diffuse passively across membranes from regions of higher to lower concentrations, and thereby yields energy. The movement of such substances against a concentration gradient is referred to as ACTIVE TRANSPORT and requires energy in the form of ATP. CHEMIOSMOSIS is a process in which the energy released when a substance moves along the gradient is used to synthesize ATP.

G. CARBOHYDRATE METABOLISM

1. During digestion, polysaccharides and disaccharides are converted to MONOSACCHARIDES, which are absorbed through the capillaries in villi and transported to the liver via the hepatic portal vein. Glucose is the principal organic substance metabolized in carbohydrate metabolism.
2. Glucose is the preferred source of energy and the fate of the absorbed glucose depends upon the body cells' energy needs.
3. If the cells require immediate energy, glucose will be oxidized by the cells. Facilitated by insulin, glucose passes into cells and becomes phosphorylated to glucose-6-phosphate.
4. Excess glucose can be stored in the liver and skeletal muscles as GLYCOGEN, converted to fat and stored in the adipose tissue, or excreted in the urine.
5. Glucose oxidation is called cellular respiration, which occurs in all cells except red blood cells and provides the cell's chief source of energy.
6. The complete oxidation of glucose to carbon dioxide and water produces a large amount of energy, and occurs in three successive steps: GLYCOLYSIS, the KREBS' CYCLE, and the ELECTRON TRANSPORT SYSTEM. GLYCOLYSIS occurs in the cytoplasm, while the KREBS' CYCLE and ELECTRON TRANSPORT SYSTEM take place in the mitochondria.
7. GLYCOLYSIS refers to the breakdown of glucose into two molecules of PYRUVIC ACID plus two molecules of ATP. If oxygen is in short supply the pyruvic acid is converted into LACTIC ACID (oxygen debt). If there are adequate supplies of oxygen, the pyruvic acid enters the Krebs' cycle.
8. PYRUVIC ACID is prepared for entrance into the Krebs' cycle by conversion to a two-carbon compound followed by the addition of a coenzyme to form ACETYL COENZYME A.
9. Each molecule of pyruvic acid that enters the Krebs' cycle produces three molecules of carbon dioxide, four molecules of NADH + H+, one molecule of reduced FAD, and GTP (guanosine triphosphate).
10. The energy originally in glucose and then in pyruvic acid is primarily the reduced coenzymes NADH + H+ and $FADH_2$.
11. The ELECTRON TRANSPORT CHAIN consists of a series of carrier molecules on the inner mitochondrial membrane that are capable of oxidation and reduction.

12. As electrons pass through the chain, there is a stepwise release of energy from the electrons for the generation of ATP.
13. In AEROBIC RESPIRATION, the terminal acceptor of the chain is molecular oxygen and the final oxidation is irreversible.
14. For every molecule of NADH+ that enters the electron transport chain, 3 ATP molecules are produced. For every molecule of $FADH_2$ that enters the electron transport chain, two ATP molecules are produced.
15. About 43% of the energy originally stored in glucose is captured by ATP; the remainder is given off as heat.
16. GLUCOSE ANABOLISM, or GLYCOGENESIS, is the conversion of glucose to glycogen for storage in the liver and skeletal muscles.
17. In the liver, GLYCOGENESIS is stimulated by insulin, and will account for the maximum storage of approximately 500 grams of glycogen.
18. The conversion of glycogen back to glucose is called GLYCOGENOLYSIS and occurs when glucose levels become low in the blood stream. Glycogenolysis is activated by glucagon and epinephrine.

H. LIPID METABOLISM

1. During digestion, TRIGLYCERIDES are broken down into FATTY ACIDS and MONOGLYCERIDES. These are carried in MICELLES for entrance into the villi, digested into GLYCEROL and FATTY ACIDS, recombined into TRIGLYCERIDES, and transported by CHYLOMICRONS through the lacteals of villi and into the thoracic duct.
2. Some fats may be oxidized to form ATP; others are stored in the adipose tissue.
3. Most lipids are used as structural molecules, or in the synthesis of essential molecules such as phospholipids, lipoproteins, thromboplastin, and cholesterol.
4. Fats are stored in adipose tissue. The adipose tissue contains lipases that catalyze the deposition of fats from chylomicrons and hydrolyze fats into fatty acids and glycerol.
5. The fats in adipose tissue are not inert; they are catabolized and mobilized constantly throughout the body.
6. Fat is generally split into FATTY ACIDS and GLYCEROL under the influence of EPINEPHRINE, NOREPINEPHRINE, and GLUCOCORTICOIDS for release from fat deposits.
7. The glycerol component can be converted into glucose by conversion into glyceraldehyde-3-phosphate via GLUCONEOGENESIS.
8. The conversion of glucose or amino acids into lipids is called LIPOGENESIS and constitutes lipid anabolism. The process is stimulated by insulin.

I. PROTEIN METABOLISM

1. During digestion, proteins are hydrolyzed into AMINO ACIDS which are absorbed by the capillaries and enter the liver via the hepatic portal vein.
2. Amino acids, under the influence of HUMAN GROWTH HORMONE and insulin, enter the body cells by active transport. Inside the cells, the amino acids are synthesized into proteins that function as enzymes, hormones, antibodies, clotting chemicals, contractile elements, and structural components.
3. Proteins may also be stored as GLYCOGEN, or used for energy.

4. Before amino acids can be catabolized, they must be converted into substances that can enter the Krebs' cycle. These conversions involve DEAMINATION, DECARBOXYLATION, and HYDROGENATION.
5. PROTEIN ANABOLISM involves the formation of peptide bonds between amino acids that produce new protein. The protein synthesis is stimulated by human growth hormone, thyroxine, and insulin.
6. Protein synthesis is carried out on the ribosomes of almost every cell in the body, directed by the cells' DNA and RNA.
7. Of the naturally occurring amino acids, 10 are referred to as essential amino acids and cannot be synthesized by the body. They must be obtained from outside sources.
8. The NONESSENTIAL AMINO ACIDS can be synthesized by the body cells by a process called TRANSAMINATION. Once the appropriate essential and nonessential amino acids are present in cells, protein synthesis occurs rapidly.

J. REGULATION OF METABOLISM

1. Based on the needs of the body, absorbed nutrients may be oxidized, stored, or converted.
2. The pathway taken by a particular nutrient is enzymatically controlled and is generally regulated by hormones.

K. MINERALS

1. MINERALS are inorganic substances that help to regulate body processes.
2. Minerals known to perform essential functions are calcium, phosphorous, sodium, chloride, potassium, magnesium, iron, sulphur, iodine, manganese, cobalt, copper, zinc, selenium, and chromium.

L. VITAMINS

1. VITAMINS are organic nutrients that maintain growth and normal metabolism. The essential function of vitamins is the regulation of physiological processes. Most vitamins serve as COENZYMES.
2. Most vitamins cannot be synthesized by the body and no single food contains all of the required vitamins.
3. On the basis of solubility, vitamins are divided into two principal groups: FAT-SOLUBLE and WATER-SOLUBLE.
4. FAT-SOLUBLE VITAMINS are absorbed with fats from the small intestine and include VITAMINS A, D, K, and E. They are generally stored in cells, particularly liver cells, so reserves can be built up.
5. WATER-SOLUBLE VITAMINS are absorbed with water from the GI tract and dissolved in body fluids. They include the B vitamins and Vitamin C. Excess quantities of these vitamins are not stored and are excreted in the urine.

M. METABOLISM AND BODY HEAT

1. A KILOCALORIE (kcal) is the amount of energy required to raise the temperature of 1000 g of water from 14.5^{o} to 15.5^{o} C.

2. The kcal is a unit of heat used to express the caloric value of foods and to measure the body's metabolic rate.
3. The apparatus used to determine the caloric value of foods is called a CALORIMETER.
4. Heat produced by the body arises from the oxidation of food. The rate at which this heat is produced is called METABOLIC RATE.
5. Metabolic rate is affected by exercise, the nervous system, hormones, body temperature, age, gender, climate, food, sleep, and malnutrition.
6. BMR is a measure of the rate at which the quiet, resting, fasting body breaks down foods and releases heat. Since thyroxine regulates the rate of food breakdown, it is also an indicator of how much thyroxine the thyroid gland is producing.
7. BMR is usually measured indirectly by measuring oxygen consumption and is expressed in kilocalories per square meter of body area per hour.
8. Most body heat produced by the oxidation of the foods we eat is transferred continuously from the body via RADIATION, CONDUCTION, CONVECTION, and EVAPORATION.
9. RADIATION is the transfer of heat from one object to another without contact.
10. CONDUCTION is the transfer of body heat to another substance or object in contact with the body.
11. CONVECTION is the transfer of body heat by the movement of air that has been warmed by the body.
12. EVAPORATION is the conversion of a liquid to a vapor. Water evaporation from the skin requires a great deal of body heat.

N. BODY TEMPERATURE REGULATION

1. Normal body temperature is maintained by a delicate balance between heat-producing and heat-losing mechanisms. If the amount of heat production equals the amount of heat loss, one maintains a constant core temperature near 38° C.
2. CORE TEMPERATURE refers to the body's temperature in body structures below the subcutaneous tissue.
3. SHELL TEMPERATURE refers to the body's temperature at the skin surface.
4. Too high a core temperature kills by denaturing body proteins, while too low a temperature causes cardiac arrhythmias that result in death.
5. Mechanisms that produce or retain heat are vasoconstriction, sympathetic stimulation, skeletal muscle contraction, thyroxine production, and perspiration. These are controlled by preoptic areas in the hypothalamus.
6. FEVER is abnormally high body temperture most probably caused by PROSTAGLANDINS with INTERLEUKIN-1.
7. The most common reason for infection is from bacteria and their toxins, and viruses. Others include heart attacks, tumors, surgery, trauma, and reactions to vaccines.
8. Up to a point, fever is beneficial in helping to fight infection and increase the rate of tissue repair during a disease. The common complications involved with a fever are dehydration, acidosis, and possible neural damage.

O. COMMON DISORDERS

1. OBESITY is defined as a body weight 10-20% above a desirable standard as a result of excessive accumulation of fat. It is implicated as a risk factor in cardiovascular disease,

hypertension, pulmonary disease, Type II diabetes mellitus, arthritis, uterus and colon cancer, varicose veins, and gallbladder disease.

2. PHENYLKETONURIA (PKU) is a genetic error of metabolism characterized by an elevation of phenylalanine in the blood.
3. CYSTIC FIBROSIS (CF) is a metabolic disease of the exocrine glands in which the absorption of Vitamins A, D, and K and of calcium are inadequate.

IV. TEACHING TIPS AND SUGGESTIONS

A. HELPFUL HINTS

1. When introducing the metabolic cycles, indicate that glycolysis, occurs in the cytoplasm, while the Krebs'' cycle and the electron transport system, both occur inside the mitochondria of body cells.
2. Refer to the relationship of the cellular respiratory chains relative to the mitochondria.
3. The essential macro- and microminerals are detailed on pages 436-437.
4. A listing of fat-soluble and water-soluble vitamins is detailed on pages 438-440.

B. ESSAYS

1. Assume that the carbohydrate component of the diet of a person is severely restricted for several days but the person continues to eat adequate amounts of fats and proteins. Describe some of the adjustments that the person's body would make to the unbalanced diet.
2. Support the statement that minerals and vitamins do not provide the body with energy. Use examples to substantiate your response.

C. TOPIC FOR DISCUSSION

1. Discuss how gluconeogenesis is important to overall metabolism.

V. AUDIOVISUAL MATERIALS

A. OVERHEAD TRANSPARENCIES

1. PAP Transparency Set (Trs. 25.1, 25.2b, 25.3, 25.4b, 25.5, 25.6, 25.7, 25.8a, 25.13 & 25.16).
2. The Table of Foods: Complete (CARO).

B. VIDEOCASSETTES

1. Metabolism: The Fire of Life (30 min.; HRM/PLP).
2. Cell Respiration: Anaerobic and Aerobic (26 min.; CFH).
3. Diet: Health and Disease (HRM).
4. Nutrition: Foods, Fads, Frauds, Facts (36 min., CFH).
5. Eating to Live (26 min.; FHS).
6. Dangerous Dieting (HRM).

C. Films: 16 mm

1. Nutrition: Fueling the Human Machine (18 min.; 1978; BFA/KSU).
2. Nutrition: What's In It For Me? (26 min.; 1975; KSU).
3. Metabolic Mazes (25 min.; 1981; MG/KSU).
4. The Human Body: Nutrition and Metabolism (14 min.; COR/KSU).
5. Food and Nutrition (10 min.; UCEMC).

D. Transparencies: 35 mm (2x2)

1. PAP Slide Set (Slides 120-122).
2. Metabolism: Structure and Regulation (42 Slides; BM/CARO).
3. ATP Functioning and Formation (K&E).
4. Dangerous Dieting, Parts I-III (228 Slides; IBIS).
5. Enzymes (46 Slides; EIL).

E. Computer Software

1. Basic Cell Respiration (Apple II; CARO).
2. Advanced Cell Respiration (Apple II; CARO).
3. Lipids (Apple II; CARO).
4. Proteins (Apple II; CARO).
5. Carbohydrates (Apple II; CARO).
6. Enzyme Investigations (Apple II; CARO).

CHAPTER 21

THE URINARY SYSTEM

CHAPTER AT A GLANCE

- KIDNEYS
- External Anatomy
- Internal Anatomy
- Blood Supply
- Nephron
- Juxtaglomerular Apparatus (JGA)
- FUNCTIONS
- *Glomerular Filtration*
- *Production of Filtrate*
- *Glomerular Filtration Rate (GFR)*
- *Regulation of GFR*
- *Tubular Reabsorption*
- *Tubular Secretion*
- HEMODIALYSIS THERAPY
- URETERS
- *Structure*
- *Functions*
- URINARY BLADDER
- *Structure*
- *Functions*
- URETHRA
- *Functions*
- URINE
- *Volume*
- *Physical Characteristics*
- *Chemical Composition*
- *Abnormal Constituents*
- COMMON DISORDERS
- MEDICAL TERMINOLOGY AND CONDITIONS
- WELLNESS FOCUS: RECURRENT UTIs

I. CHAPTER SYNOPSIS

This chapter is a study of the organs of the urinary system and their role in urine production, urine excretion, and the maintenance of the composition and volume of blood. The macroscopic structure and function of the kidney is followed by the structure and function of the microscopic nephron units. Attention is given to the physiological processes involved in the production of urine at the nephron level with careful consideration paid to filtration, reabsorption and secretion processes. The function of the kidney in maintaining pH and blood pressure is also considered. The structure, histology, and physiology of the ureters, urinary bladder, and urethra are also detailed. The chapter concludes with the physical characteristics of urine, common disorders, medical terms, and conditions associated with the urinary system.

II. LEARNING GOALS/STUDENT OBJECTIVES

1. Describe the structure and blood supply to the kidneys.
2. Describe how the kidneys filter the blood and regulate its volume, chemical composition, and pH.
3. Describe the principles and importance of hemodialysis.
4. Describe the structure and functions of the ureters.
5. Describe the structure and functions of the urinary bladder.
6. Describe the structure and functions of the urethra.
7. Describe the normal and abnormal components of urine.

III. SAMPLE LECTURE OUTLINE

A. Introduction

1. The primary functions of the urinary system are to:
   - Regulate the concentration and volume of blood.
   - Regulate blood pressure.
   - Contribute to metabolism.
2. The organs of the urinary system include the kidneys, ureters, urinary bladder, and urethra.

B. External Kidney Anatomy

1. The kidneys are located retroperitoneally, and are attached to the posterior abdominal wall.
2. Near the center of the concave medial border of the kidney, is a notch called the hilus through which the ureter leaves and the blood vessels, lymphatic vessels, and nerves enter and exit.
3. Three layers of tissue surround each kidney: the innermost renal capsule, the adipose capsule, and the outer renal fascia.

C. Internal Kidney Anatomy

1. Internally, the kidneys consist of a CORTEX, MEDULLA, PYRAMIDS, PAPILLAE, COLUMNS, CALYCES, and a PELVIS.
2. The functional unit of the kidney is the NEPHRON, which consists of two major portions: RENAL TUBULES called the GLOMERULAR or BOWMAN'S CAPSULE, and a network of CAPILLARIES called the GLOMERULUS.
3. The BOWMAN'S CAPSULE surrounds the glomerular capillary network. Collectively, the Bowman's capsule and capillary network are referred to as the RENAL CORPUSCLE.
4. The inner wall of the glomerular capsule and glomerulus together form a filtering membrane called the ENDOTHELIAL-CAPSULAR MEMBRANE, which permits the passage of solutes and fluid from the blood into the nephron.
5. From the endothelial-capsular membrane the fluid and solutes pass into the PROXIMAL CONVOLUTED TUBULE and into the DESCENDING LIMB OF THE LOOP OF the NEPHRON. Next, the fluid and solutes pass through the LOOP OF HENLE and up the ASCENDING LIMB of the LOOP of the NEPHRON to the DISTAL CONVOLUTED TUBULE.
6. The DISTAL CONVOLUTED TUBULE terminates by emptying into the COLLECTING TUBULE. In the medulla of the kidneys, the collecting tubules unite to form the PAPILLARY DUCT, which transmits the fluid and solutes into the CALYCES and into the RENAL PELVIS of the kidney.
7. The JUXTAGLOMERULAR APPARATUS (JGA) consists of juxtaglomerular cells and the MACULA DENSA of the distal convoluted tubule. The JGA helps to regulate renal blood pressure.

D. Kidney Function

1. NEPHRONS are the functional units of the kidneys. They carry out three important functions: the control of blood concentration by the removal of selected amounts of water and solutes, the regulation of pH, and the removal of toxic wastes from the blood.
2. The nephrons form urine by three principal processes: GLOMERULAR FILTRATION, TUBULAR REABSORPTION, and TUBULAR SECRETION.

E. Nephron Filtration

1. The force of fluids and dissolved substances through a membrane by pressure occurs in the renal corpuscles of the kidney across the endothelial-capsular membrane.
2. When blood enters the glomerulus, the blood pressure (GLOMERULAR HYDROSTATIC BLOOD PRESSURE) forces water and dissolved components through the endothelial pores of the capillaries, through the BASEMENT MEMBRANE, and on through the filtration slits of the cells that make up the BOWMAN'S CAPSULE.
3. FILTRATION of blood depends on the force of glomerular hydrostatic blood pressure in relation to two opposing forces: capsular hydrostatic pressure and blood colloidal osmotic pressure. This relationship is known as the NET FILTRATION PRESSURE (NFP).
4. The volume of filtrate that flows out of all the renal corpuscles per unit time is called GLOMERULAR FILTRATION RATE (GFR), and is approximately 125 ml per min.

F. Tubular Reabsorption

1. Tubular reabsorption is the movement of certain components of the filtrate back into the blood. In this way, the substances needed by the body, including glucose, water, amino acids, and ions, are retained.
2. The maximum amount of any substance that can be reabsorbed under any conditions is called the tubular maximum (TM).
3. Water reabsorption is driven by sodium transport. Approximately 80% of the reabsorbed water is returned by obligatory water reabsorption. Passage of the remaining water in the filtrate can be regulated by ADH. If the reabsorption of water is controlled by ADH, it is referred to as facultative water reabsorption. This is the major mechanism for controlling the water content of the blood.

G. Tubular Secretion

1. Chemicals not needed by the body, such as ions, nitrogenous wastes, and certain drugs, are discharged into the urine by tubular secretion.
2. The kidneys help to maintain blood pH by secreting hydrogen ions, and by increasing or decreasing bicarbonate concentration.

H. Hemodialysis Therapy

1. The filtration of blood by artificial means is called hemodialysis.
2. The artificial kidney machine filters the blood of wastes and adds nutrients.

I. Homeostasis

1. The kidneys assume the major role in excretion, but the lungs, integument, and gastrointestinal tract also assume excretory functions.

J. Ureters

1. The ureters are retroperitoneal tubes consisting of an innermost mucosa, middle muscularis, and an outer fibrous coat.

2. The ureters transport urine from the renal pelvis of the kidneys to the urinary bladder primarily by peristalsis. Hydrostatic pressure and gravity also contribute to the urine flow in the ureters.

K. Urinary Bladder

1. The urinary bladder is located posterior to the symphysis pubis in the pelvic cavity. Its function is to store urine prior to micturition.
2. At the base of the urinary bladder is a small, smooth, triangular area called the trigone. The ureters enter the urinary bladder near the two posterior points of the triangle. The urethra drains the bladder from the anterior point of the triangle.
3. Histologically, the urinary bladder consists of an innermost mucosa with rugae, a middle muscularis, and an outer serous coat.

4. In the area around the opening to the urethra, the circular fibers of the muscularis form the INTERNAL URETHRAL SPHINCTER.
5. Deep in the internal sphincter is the EXTERNAL SPHINCTER, which is composed of voluntary skeletal muscle.
6. Urine is expelled from the bladder by a process called micturition.

L. URETHRA

1. The URETHRA is a tube leading from the floor of the urinary bladder to the exterior of the body.
2. Its function is to RELEASE THE URINE from the body. In the male, the urethra also serves as a passageway for reproductive fluid to be discharged from the body.

M. URINE

1. The by-product of kidney activity is URINE. In a healthy individual, its volume, pH, and solute concentration vary with the needs of the internal environment.
2. In certain pathological conditions the characteristics of urine may change. An analysis of the volume and physical, chemical, and microscopic properties of urine reveals a great deal about the physiological state of the body.
3. The VOLUME of urine is influenced by blood pressure, blood concentration, diet, temperature, diuretics, mental state, and general health.
4. The PHYSICAL CHARACTERISTICS of urine generally evaluated by urinalysis are color, turbidity, odor, pH, and specific gravity.
5. CHEMICALLY, normal urine consists of approximately 95% water and 5% solutes. The solutes include urea, creatinine, uric acid, hippuric acid, ketone bodies, salts, and ions.

N. URINARY DISORDERS

1. GLOMERULONEPHRITIS is an inflammation of the glomeruli of the kidney, generally from an allergic reaction to the toxins given off by streptococci bacteria.
2. PYELITIS is an inflammation of the renal pelvis and its calyces.
3. CYSTITIS is an inflammation of the urinary bladder, usually caused by a bacterial infection, chemical irritation, or mechanical injury.
4. NEPHROTIC SYNDROME is characterized by protein in the urine due to increased endothelial-capsular membrane permeability. It results from a variety of diseases, drugs, heavy metals, and hypertension.
5. RENAL FAILURE is the reduction, cessation of glomerular filtration, and the prevention of metabolic waste elimination from the body, and can be fatal.

IV. TEACHING TIPS AND SUGGESTIONS

A. HELPFUL HINTS

1. A detailed macroscopic dissection of a pig kidney, or other large mammalian kidney, is useful.
2. Microscopic examination of the renal corpuscular units is helpful in understanding their relationship to one another.

3. It is helpful to compare the constituents of plasma, glomerular filtrate, and urine to observe similarities and dissimilarities.

B. Essays

1. Trace a drop of blood from its entrance into the renal artery to its exit out of the renal vein as it passes through the kidney. Name in sequence, the vessels along the way, and explain what happens during glomerular filtration, tubular reabsorption, and secretion.
2. Trace a drop of glomerular filtrate from its origin to its exit at the urethra. Be sure to include all ducts, tubes, and passageways.

V. AUDIOVISUAL MATERIALS

A. Overhead Transparencies

1. PAP Transparency Set (Trs. 26.1, 26.4a, 26.5b, 26.6a&b, 26.8a&b, 26.9a&b, 26.10, 26-13a-c, 26.14-26.16, 26.17a-c, 26.18-26.20 & 26.23a).
2. Function and Excretion of Kidneys (CARO).
3. Human Urinary System (CARO).

B. Videocassettes

1. Water (26 min.; FHS).
2. Excretion (29 min.; C; Sd; 1978; IM).
3. Kidney Disease (26 min.; C; Sd; 1992; FHS).
4. The Mammalian Kidney (EIL).
5. Water (26 min.; C; Sd; 1990; FHS).
6. Work of the Kidneys (23 min.; C; Sd; 1989; GAF).

C. Films: 16 mm

1. The Human Body: Excretory System (14 min.; 1980; COR/KSU).
2. The Work of the Kidneys (20 min.; 1972; EBEC/KSU).
3. Kidney Function in Health (38 min.; LILLY).
4. Excretion (28 min.; 1960; McG/KSU).
5. Elimination (11 min.; UWF).
6. I Am Joe's Kidney (30 min.; 1984; PYR/KSU).

D. Transparencies: 35 mm (2x2)

1. PAP Slide Set (Slides 123-129).
2. AHA Slide Set.
3. Urinary System- Unit 9 (7 Slides; RJB).
4. The Mammalian Kidney (72 Slides; (EIL).
5. Urinary System and Its Function (20 Slides; EIL).
6. The Urinary System Set (CARO).

E. Computer Software

1. The Kidney: Structure and Functions (Apple; EIL).
2. Urinary System Probe (Apple; PLP).
3. Urinary System (Apple; PLP).
4. The Kidney (Apple; PLP).
5. Body Language: Urinary System (Apple II Series; ESP).

CHAPTER 22

FLUID, ELECTROLYTE, AND ACID-BASE BALANCE

CHAPTER AT A GLANCE

- FLUID COMPARTMENTS AND FLUID BALANCE
- WATER
    - *Fluid Intake, Gain, and Output*
    - *Regulation of Fluid Intake*
    - *Regulation of Fluid Output*
- ELECTROLYTES
    - *Concentration of Electrolytes*
    - *Distribution of Electrolytes*
    - *Functions and Regulation*
        - *Sodium*
        - *Potassium*
        - *Calcium*
        - *Magnesium*
        - *Chloride*
        - *Phosphate*
- MOVEMENT OF BODY FLUIDS
    - *Between Plasma and Interstitial Compartments*
    - *Between Interstitial and Intracellular Compartments*
- ACID-BASE BALANCE
    - *Buffer Systems*
        - *Carbonic Acid-Bicarbonate Buffer System*
        - *Phosphate Buffer System*
        - *Hemoglobin Buffer System*
        - *Protein Buffer System*
    - *Respiration*
    - *Kidney Excretion*
- ACID-BASE IMBALANCES
- WELLNESS FOCUS: PROLONGED PHYSICAL ACTIVITY- A CHALLENGE TO FLUID AND ELECTROLYTE BALANCE

I. CHAPTER SYNOPSIS

This chapter outlines the fluid present in the body, the regulation of fluid intake and output, the electrolyte composition and concentration, and the distribution of the fluid compartments. Attention is given to the function of electrolytes and how their levels are controlled, and factors that govern fluid between compartments. The major ions and their functions are detailed. Consideration is given to the movement of body fluids between plasma and intercellular compartments, and intercellular and intracellular compartments. The maintenance of pH of body fluids through the effects of buffers, respirations, and kidney excretion are discussed. The chapter concludes with a discussion of acid-base balance.

II. LEARNING GOALS/STUDENT OBJECTIVES

1. Explain the routes of fluid intake and output, and explain how intake and output are regulated.
2. Describe the general functions of electrolytes and how they are distributed.
3. Discuss the functions and regulation of sodium, potassium, calcium, magnesium, chloride, and phosphate.
4. Describe how fluids move between compartments.
5. Explain how buffers, respiration, and kidney excretion help maintain pH.

III. SAMPLE LECTURE OUTLINE

A. WATER

1. WATER is the most abundant substance in the body and represents 45% to 75% of the body's weight, depending upon the amount of fat present, gender, and age.
2. FLUID INTAKE generally equals fluid output, so that a constant body fluid volume is maintained.
3. The principal sources of fluid intake include ingested fluids and foods, and water produced by catabolism.
4. Fluid leaves the body from the kidneys, skin, lungs, the gastrointestinal and urinary tracts.
5. The stimulus for fluid intake is dehydration resulting from thirst sensations. The fluid output is homeostatically adjusted via antidiuretic hormones and aldosterone, which regulate kidney urine output.

B. FLUID COMPARTMENTS AND FLUID BALANCE

1. Body fluid consists of water and its dissolved solutes. Approximately two-thirds of the body's fluid is located in the cells and is called INTRACELLULAR FLUID.
2. The remaining third lies outside of the cells and is called EXTRACELLULAR FLUID. It includes interstitial fluid, plasma, lymph, cerebrospinal fluid, gastrointestinal fluid, synovial fluid, fluids of the eyes, pleural, pericardial and peritoneal fluid, and glomerular filtrate in the kidneys.
3. FLUID BALANCE refers to the distribution of fluids in response to certain body requirements.
4. OSMOSIS is the principal mechanism used to move water into and out of the body compartments, and the concentration of solutes in the water is the major determinant of the fluid balances.

5. Most of the solutes in the fluids are electrolytes, or substances that are dissociated into ions.
6. Fluid balance generally refers to water balance, but also indirectly refers to electrolyte balance. The two balances are inseparable.

C. REGULATION OF FLUID INTAKE

1. THIRST is the principal regulator of fluid intake. Thirst operates in the following sequence:
   ∞ When water loss exceeds water gain DEHYDRATION results.
   ∞ Dehydration stimulates thirst by:
   - decreasing the production of saliva.
   - increasing blood osmotic pressure.
   - decreasing blood volume.
2. A decrease in saliva production causes dryness of the mucosa. When blood osmotic pressure increases hypothalamic osmoreceptors are stimulated. When blood volume decreases, the juxtaglomerular apparatus is stimulated to promote the synthesis of ANGIOTENSIN II.
3. The combined effects of dry mouth, angiotensin II, and stimulation of the osmoreceptors, stimulates the thirst center in the hypothalamus.
4. The hypothalamus increases the sensation for thirst.
5. Thirst results in fluid intake until appropriate volumes are restored.

D. ELECTROLYTES

1. NONELECTROLYTES and ELECTROLYTES are dissolved chemicals in the body fluids.
2. NONELECTROLYTES are compounds with covalent bonds and include most organic compounds, such as glucose, urea, and creatine.
3. ELECTROLYTES are chemicals that dissolve in body fluids and dissociate either as cations (positive ions) or anions (negative ions). The major classes are ACIDS, BASES, and SALTS.
4. ELECTROLYTES serve three basic functions in the body: they are essential minerals, they control osmosis of water between body compartments, and they maintain the acid-base balance required for normal cellular function.
5. Electrolyte concentration is expressed in milliequivalents per liter (mEq/l).

E. FUNCTION AND REGULATION OF IONS

1. The plasma, interstitial fluid, and intracellular fluid contain varying kinds and amounts of electrolytes.
2. SODIUM (Na^+) is the most abundant extracellular ion, representing approximately 90% of extracellular cations. It is involved in impulse transmission and muscle contraction, and participates in fluid and electrolyte balance by creating most of the osmotic pressure of the extracellular fluid.
3. The average daily intake of sodium far exceeds the body's daily normal requirements. Excess sodium is excreted by the kidneys. During periods of reduced sodium intake, the kidneys conserve sodium.
4. The sodium level in the blood is hormonally controlled by ALDOSTERONE, secreted by the adrenal cortex.
5. CHLORIDE (Cl^-) is the major extracellular anion, and assumes a role in regulating osmotic pressure and forming HCl in the GI tract.

6. The REGULATION of chloride ions is indirectly controlled by ALDOSTERONE via the reabsorption of sodium. Chloride is passively reabsorbed with sodium.
7. POTASSIUM (K^+) is the most abundant cation in intracellular fluid and is involved in maintaining fluid volume, impulse conduction, muscle contraction, and the regulation of pH. Potassium level is controlled by ALDOSTERONE.
8. CALCIUM (Ca^{2+}) is the most abundant cation in the body. It is principally an extracellular ion, but is also a component of bones and teeth. It functions in blood clotting, neurotransmitter release, neuromuscular conduction for muscle contraction, maintenance of muscle tone, and the excitability of muscle tissue. The level of calcium in the blood is controlled principally by PARATHYROID HORMONE and CALCITONIN.
9. PHOSPHATE (HPO_4^{2-}) is an intracellular ion that is a structural component of bones and teeth. It is also required for the synthesis of nucleic acids and high-energy compounds such as ATP, and for buffer reactions. Phosphate levels are controlled by PARATHYROID HORMONE and CALCITONIN.
10. MAGNESIUM (Mg^{2+}) is an intracellular electrolyte, and functions by activating several enzyme systems. It also functions in triggering the sodium potassium pump, and preserves the structure of DNA, RNA, and ribosomes. It aids in neuromuscular activities, CNS neural transmission, and myocardial functioning. Magnesium levels are controlled by ALDOSTERONE.

F. MOVEMENT OF BODY FLUIDS

1. Movement of body fluids between the plasma and interstitial compartment occurs across CAPILLARY MEMBRANES.
2. The movement of water and non-protein solutes through capillary walls, occurs primarily by DIFFUSION.
3. HYDROSTATIC PRESSURE is due to the presence of water in the fluids. Blood pressure in the capillaries, called BLOOD HYDROSTATIC PRESSURE (BHP), moves fluid out of the capillaries into the interstitial fluid.
4. INTERSTITIAL FLUID HYDROSTATIC PRESSURE (IFHP) opposes BHP and moves fluid from the interstitial spaces into the capillaries.
5. OSMOTIC PRESSURES are due to the presence of proteins that are too large to diffuse in the blood and interstitial fluid.
6. BLOOD OSMOTIC PRESSURE (BOP) moves fluid from the interstitial spaces into the capillaries. INTERSTITIAL FLUID OSMOTIC PRESSURE (IFOP) opposes BOP and moves fluid out of the capillaries into the interstitial fluid.
7. The difference between the two fluids that move fluid out of the plasma and the two fluids that push fluid into the plasma is called the NET FILTRATION PRESSURE (NFP), and can be represented by: NFP = (BHP = IFOP) - (IFHP + BOP).
8. INTRACELLULAR FLUID has a higher osmotic pressure than interstitial fluid. The principal cation inside the cell is potassium, whereas the principal cation outside the cell is sodium.
9. Fluid imbalances between the intracellular fluid and interstitial fluid are generally due to changes in the sodium and potassium concentrations. These concentrations are regulated by the secretion of antidiuretic hormone and aldosterone.

G. ACID-BASE BALANCE

1. Electrolytes not only regulate water movement, but also help regulate the body's ACID-BASE BALANCE.
2. The overall ACID-BASE BALANCE depends upon the hydrogen ion concentration of body fluids, particularly in the extracellular fluid.
3. HYDROGEN IONS are produced as a result of the metabolic activities of the cell and form carbonic acid and bicarbonate ions as follows: $CO_2 + H_2O \longrightarrow H_2CO_3 \longrightarrow H^+ + HCO_3^-$.
4. Human buffer systems consist of a WEAK ACID and a WEAK BASE that function to prevent drastic changes in the pH of the body fluid by rapidly changing strong acids and bases into weak acids and bases.
5. STRONG ACIDS dissociate into hydrogen ions more easily than weak acids and thus, lower pH more than weak acids.
6. Similarly, STRONG BASES raise pH more than weak bases because strong bases dissociate into hydroxide (OH^-) ions.
7. The principal buffer systems in the body are the CARBONIC ACID-BICARBONATE SYSTEM, the PHOSPHATE SYSTEM, the HEMOGLOBIN-OXYHEMOGLOBIN SYSTEM, and the PROTEIN SYSTEM.

H. CARBONIC ACID-BICARBONATE BUFFER SYSTEM

1. The CARBONIC ACID-BICARBONATE BUFFER SYSTEM is an important regulator of blood pH. It contributes bicarbonate ions to buffer strong acids.
2. Strong acids such as hydrochloric acid can be buffered by SODIUM BICARBONATE to produce a weak acid (carbonic acid) and salt (NaCl).
3. Strong bases, such as sodium hydroxide, can be buffered by carbonic acid to produce sodium bicarbonate and water.

I. PHOSPHATE BUFFER SYSTEM

1. The PHOSPHATE BUFFER SYSTEM helps regulate pH in red blood cells and especially in the kidney tubular fluids. The phosphate system helps the kidneys maintain normal blood pH by acidification of the urine.
2. Two phosphate buffers, SODIUM MONOHYDROGEN PHOSPHATE and SODIUM DIHYDROGEN PHOSPHATE, buffer strong acids and bases respectively.

J. HEMOGLOBIN BUFFER SYSTEM

1. The HEMOGLOBIN BUFFER SYSTEM buffers carbonic acids in the blood. At the venous end, carbon dioxide moves into the red blood cells and combines with water to form carbonic acid.
2. As OXYHEMOGLOBIN gives up oxygen in the tissues, some of the hemoglobin becomes reduced and attracts hydrogen ions from carbonic acids, reducing it to a weak acid.

K. PROTEIN BUFFER SYSTEM

1. The PROTEIN BUFFER SYSTEM is most abundant in the body cells and plasma.

2. The amino acids in proteins contain at least one carboxyl group and one amine group. In solution, the carboxyl group ionizes in water and buffers bases to form water and carboxyl ions.
3. In the presence of excess hydrogen ions, free carboxyl ions will recombine with the hydrogen ions and form carboxyl, thereby raising pH.
4. The amine group accepts hydrogen ions to form ammonium ions, which are excreted from body fluids in urine.

L. ACID-BASE IMBALANCES

1. In ACIDOSIS, blood pH ranges from 7.35 to 6.80 or lower. In ALKALOSIS, the pH range is 7.45 to 8.00, or higher.
2. COMPENSATION refers to the physiological responses associated with acid-base imbalances.
3. ACIDOSIS depresses the central nervous system, causing disorientation and coma. ALKALOSIS can over-excite the central nervous system, causing nervousness, muscular spasm, and convulsions.
4. RESPIRATORY ACIDOSIS is caused by hypoventilation while respiratory alkalosis is causes by hyperventilation.
5. METABOLIC ACIDOSIS is caused by an abnormal increase in acidic metabolic end-products other than carbon dioxide.

IV. TEACHING TIPS AND SUGGESTIONS

A. HELPFUL HINTS

1. A tabulated comparison of acidosis and alkalosis is helpful.
2. It is useful to have the students draw a schematic diagram of the mechanics of pH and buffer controls.

B. ESSAYS

1. Given the following, calculate net filtration pressure.

   Blood hydrostatic pressure = 24 mm Hg
   Interstitial fluid osmotic pressure = 5 mm Hg
   Interstitial fluid hydrostatic pressure = 4 mm Hg
   Blood osmotic pressure = 25 mm Hg

2. Describe the mechanism by which a sodium deficit in interstitial fluid may lead to water intoxification and shock.

V. AUDIOVISUAL MATERIALS

A. Overhead Transparencies

1. PAP Transparency Set. (Trs. 27.3-27.6).

B. Videocassettes

1. Acids, Bases, and Salts (20 min.; 1983; COR/KSU).
2. Buffers (EIL).
3. Water (26 min.; C; Sd; 1990; FHS).

C. Films: 16 mm

1. Diffusion and Osmosis (13 min.; 1973; EBEC/KSU).
2. Water in Biology (21 min.; 1967; FA/KSU).
3. Dynamics of Fluid Exchange (39 min.; ACCI).

D. Transparencies: 35 mm (2x2)

1. PAP Slide Set (Slide 130).
2. Buffers Set (64 Slides; BM/CARO).
3. pH, Osmosis, and Diffusion (30 Slides; CARO).
4. Homeostasis (150 Slides with cassettes; CARO).

E. Computer Software

1. Osmosis (Apple II; CARO).
2. Osmotic Pressure (Apple II; CARO).

CHAPTER 23

THE REPRODUCTIVE SYSTEMS

CHAPTER AT A GLANCE

- MALE REPRODUCTIVE SYSTEM
- SCROTUM
- TESTES
- *Spermatogenesis*
- *Sperm*
- *Testosterone and Inhibin*
- MALE PUBERTY
- DUCTS
- *Ducts of the Testes*
- *Epididymis*
- *Ductus (Vas) Deferens*
- *Ejaculatory Duct*
- *Urethra*
- ACCESSORY SEX GLANDS
- *Semen*
- *Penis*

- FEMALE REPRODUCTIVE SYSTEM
- OVARIES
- *Oogenesis*
- UTERINE (FALLOPIAN) TUBES
- UTERUS
- VAGINA
- VULVA
- PERINEUM
- MAMMARY GLANDS
- FEMALE PUBERTY
- FEMALE REPRODUCTIVE CYCLE (FRC)
- *Hormonal Regulation*
- *Menstrual Phase (Menstruation)*
- *Preovulatory Phase*
- *Ovulation*
- *Postovulatory Phase*
- *Menopause*
- COMMON DISORDERS
- MEDICAL TERMINOLOGY AND CONDITIONS
- WELLNESS FOCUS: FERTILITY FRAGILITY

I. CHAPTER SYNOPSIS

Students are introduced to the structure and function of the male and female organs of reproduction. Attention is given to the process of spermatogenesis, structure of spermatozoa, and the influence of testosterone and inhibin on sperm production. Detail is then given to the numerous ducts in the male reproductive tract including the epididymis, vas deferens, ejaculatory duct, and urethra. The structure and function of the seminal vesicles, prostate gland and bulbo-urethral glands as accessory structures for the production of seminal fluid is then detailed. The composition of semen and the structure of the penis are then discussed. The chapter continues with a discussion of the ovaries and the process of oogenesis in the female reproductive system. The structure and function of the other female organs of reproduction are detailed, followed by a discussion of the perineum, mammary glands and female pubescence. Considerable attention is then given to the female reproductive cycle, especially hormonal regulation and the normal phases of the menstrual cycle. The chapter concludes with a discussion of menopause, common disorders, medical terms, and conditions associated with the reproductive systems.

II. LEARNING GOALS/STUDENT OBJECTIVES

1. Describe the structure and functions of the male reproductive system.
2. Describe how sperm cells are produced.
3. Explain the functions of the male reproductive hormones.
4. Describe the structure and functions of the female reproductive organs.
5. Describe how ova are produced.
6. Explain the functions of the female reproductive hormones, and then define the menstrual and ovarian cycles and explain how they are related.

III. SAMPLE LECTURE OUTLINE

MALE REPRODUCTIVE SYSTEM

A. INTRODUCTION

1. REPRODUCTION is the process by which genetic material is passed on from one generation to the next.
2. The organs of reproduction are classified as GONADS, DUCTS, and ACCESSORY GLANDS.
3. The MALE structures of reproduction include the TESTES, EPIDIDYMIS, VAS DEFERENS, EJACULATORY DUCT, URETHRA, SEMINAL VESICLES, PROSTATE GLAND, BULBO-URETHRAL GLANDS, and PENIS.

B. SCROTUM

1. The SCROTUM is a cutaneous outpouching of the abdomen that supports the testes.
2. The production and survival of the spermatozoa requires a temperature LOWER than body temperature. The temperature of the testes is regulated by the CREMASTER muscle, which elevates them and brings them closer to the pelvic cavity, or relaxes them causing the testes to move away from the pelvic cavity.

C. TESTES

1. The TESTES are paired, oval-shaped glands (gonads) located in the SCROTUM.
2. The testes develop in the embryo's posterior abdominal wall and usually begin their descent to the scrotum through the inguinal canals during the latter half of the seventh month of development.
3. The testes contain SEMINIFEROUS TUBULES in which sperm are produced by the process of SPERMATOGENESIS. The testes also contain SUSTENTACULAR (SERTOLI) CELLS that nourish sperm cells and secrete INHIBIN, and INTERSTITIAL CELLS OF LEYDIG, which produce the male hormone TESTOSTERONE.
4. SPERMATOGENESIS produces HAPLOID (n) spermatozoa, and involves several phases including meiosis and mitosis.
5. Ova and sperm are collectively called GAMETES, or SEX CELLS, and are produced by the gonads.
6. Uninucleate somatic cells divide by mitosis; each daughter cell receives a full complement of chromosomes (46) and is said to be diploid (2n).
7. Immature gametes divide by MEIOSIS, in which the pairs of chromosomes are split so that the mature gamete has only 23 chromosomes, and is said to be haploid (n).
8. PUBERTY refers to the period of time when the SECONDARY SEX CHARACTERISTICS begin to develop and their potential for sexual reproduction is reached. Male puberty begins at an average age of 10-11 and ends at an average of 15 to 17 years of age. Its onset is signaled by increased levels of LH, FSH, and TESTOSTERONE.
9. FSH and LH, secreted by the ANTERIOR PITUITARY under the influence of GONADOTROPIN RELEASING HORMONE (GnRH), initiate spermatogenesis by stimulating the production of testosterone.
10. SPERMATOGENESIS occurs in the testes and forms four haploid spermatozoa from each immature spermatocyte that undergoes the meiotic process.
11. SPERMATOGENESIS is the process in which immature spermatogonia develop into mature spermatozoa. The spermatogenesis sequence includes reduction division (MEIOSIS I), equatorial division (MEIOSIS II), and spermatogenesis.
12. MATURE SPERMATOZOA consists of a HEAD, MIDPIECE, and TAIL. They are produced at a rate of 300 million per day, and once ejaculated, have a life expectancy of 48 hours.
13. TESTOSTERONE controls growth, development, and maintenance of sex organs, stimulates bone growth, protein anabolism, and sperm maturation, and stimulates the development of the male secondary sex characteristics.
14. INHIBIN is produced by the Sertoli cells and inhibits FSH to regulate the rate of spermatogenesis.

D. DUCTS

1. The duct system of the testes includes the SEMINIFEROUS TUBULES, STRAIGHT TUBULES, and RETE TESTIS.
2. Sperm are transported out of the testes through the EFFERENT DUCTS and stored in the EPIDIDYMIS, which is lined with STEREOCILIA and is the site for sperm maturation and storage.
3. The VAS DEFERENS stores sperm and propels them toward the urethra during EJACULATION.
4. The SPERMATIC CORD is a supporting structure of the male reproductive system and consists of the VAS DEFERENS, the TESTICULAR ARTERY, AUTONOMIC NERVES, VEINS that drain the testes, LYMPHATICS, and the CREMASTER MUSCLE.

5. The EJACULATORY DUCTS are formed by the union of the ducts from the SEMINAL VESICLES and the VAS DEFERENS. Their function is to eject sperm into the prostatic urethra.
6. The male URETHRA is subdivided into three portions: PROSTATIC, MEMBRANOUS, and SPONGY (CAVERNOUS).

E. ACCESSORY SEX GLANDS

1. The SEMINAL VESICLES secrete an alkaline, viscous fluid that constitutes about 60% of the volume of semen and contributes to sperm viability.
2. The PROSTATE GLAND secretes a slightly acidic fluid that constitutes about 15-33% of the volume of semen and contributes to sperm motility.
3. The BULBO-URETHRAL (COWPER'S) GLANDS secrete mucus for lubrication and a substance that neutralizes acid.

F. SEMEN

1. SEMEN, or SEMINAL FLUID, is a mixture of spermatozoa and accessory sex gland secretions that provides nutrients and the fluid in which the spermatozoa are transported. It also neutralizes the acidity of the male urethra and female vagina.
2. SEMEN also contains SEMINALPLASMIN, an antibiotic, and prostatic enzymes which coagulate, then liquefy semen to aid in its movement through the cervix of the uterus.

G. PENIS

1. The PENIS is the male organ of COPULATION that consists of a root, body, and glans penis. It is used to introduce spermatozoa into the vagina.
2. Expansion of its blood sinuses under the influence of sexual excitation causes an ERECTION.
3. CIRCUMCISION is a surgical procedure in which part of or all of the PREPUCE (foreskin) is removed.

FEMALE REPRODUCTIVE SYSTEM

A. INTRODUCTION

1. The FEMALE ORGANS of reproduction include the OVARIES (GONADS), UTERINE (FALLOPIAN) TUBES, UTERUS, VAGINA, and VULVA.
2. The MAMMARY GLANDS are also considered part of the reproductive system.

B. OVARIES

1. The OVARIES are female gonads which are located in the upper pelvic cavity, on either side of the uterus. They are maintained in position by a series of ligaments.
2. The ovary produces SECONDARY OOCYTES, discharges secondary oocytes at ovulation, and secretes estrogens, progesterone, relaxin, and inhibin.
3. OOGENESIS occurs in the ovaries and results in the formation of a single haploid (n) secondary oocyte.

4. The OOGENESIS SEQUENCE includes a reduction division (MEIOSIS I), equatorial division (MEIOSIS II), and maturation.

C. UTERINE (FALLOPIAN) TUBES

1. The UTERINE (FALLOPIAN) TUBES transport the ova from the ovaries to the uterus, and are the normal site of fertilization.
2. Ciliated cells and peristaltic contractions help move the secondary oocyte through the uterine tubes towards the uterus.

D. UTERUS

1. The UTERUS (womb) is an inverted pear-shaped organ that functions in TRANSPORTING SPERMATOZOA, MENSTRUATION, IMPLANTATION OF A FERTILIZED OVUM, DEVELOPMENT OF THE FETUS DURING PREGNANCY, and LABOR.
2. The uterus is normally held in position by a series of ligaments, and histologically consists of an outer PERIMETRIUM, middle MYOMETRIUM, and inner ENDOMETRIUM.
3. Blood is supplied to the uterus by the uterine arteries.

E. VAGINA

1. The VAGINA is the passageway for spermatozoa and menstrual flow, the receptacle for the penis during sexual intercourse, and the lower portion of the birth canal.
2. The VAGINA is capable of considerable distention to accommodate its functions.

F. VULVA

1. The VULVA, or pudendum, is a collective term for the EXTERNAL GENITALS of the female.
2. It consists of the MONS PUBIS, LABIA MAJORA and MINORA, CLITORIS, VESTIBULE, VAGINAL and URETHRAL ORIFICES, HYMEN, BULB OF THE VESTIBULE, and PARAURETHRAL, GREATER VESTIBULAR, and LESSER VESTIBULAR GLANDS.

G. PERINEUM

1. The PERINEUM is a diamond-shaped area between the vaginal orifice and anal opening in the female.
2. It is of clinical importance in the female because it is generally incised during vaginal deliveries to accommodate the fetal head in a procedure called an EPISIOTOMY.

H. MAMMARY GLANDS

1. The MAMMARY GLANDS are modified SUDORIFEROUS GLANDS.
2. Milk-secreting cells, referred to as alveoli, are clustered in small lobules within the breasts.
3. Mammary gland development is dependent upon ESTROGENS and PROGESTERONE produced by the ovaries.
4. The ESSENTIAL FUNCTION of the mammary glands is LACTATION, the secretion and ejection of milk.

5. MILK SECRETION is due to the hormone PROLACTIN (PRL), and MILK EJECTION is stimulated by OXYTOCIN (OT), which is released by the posterior pituitary gland in response to sucking.

I. FEMALE REPRODUCTIVE CYCLE (FRC)

1. The function of the MENSTRUAL CYCLE is to prepare the endometrium each month for the reception of a fertilized egg.
2. The menstrual and ovarian cycles are controlled by GONADOTROPIN RELEASING HORMONE (GnRH), which stimulates the release of FOLLICLE STIMULATING HORMONE (FSH) and LEUTINIZING HORMONE (LH).
3. FSH stimulates the initial development of the OVARIAN FOLLICLES and the secretion of estrogens by the ovaries.
4. LH stimulates further development of the OVARIAN FOLLICLES, OVULATION, and the secretion of ESTROGENS and PROGESTERONE by the ovaries.
5. ESTROGENS stimulate growth, development, and maintenance of female reproductive structures, the development of the secondary sexual characteristics, regulate fluid and electrolyte balance, and stimulate protein anabolism.
6. PROGESTERONE (PROG) works with estrogens to prepare the endometrium for implantation and the mammary glands for milk production.
7. INHIBIN may be important in decreasing secretion of FSH and LH toward the end of the menstrual cycle. Relaxin relaxes the symphysis pubis and helps dilate the uterine cervix to facilitate delivery.
8. The events occurring during the menstrual cycle are divided into four phases: the MENSTRUAL PHASE, PREOVULATORY PHASE, OVULATION, and POSTOVULATORY PHASE.

J. FRC: MENSTRUAL PHASE

1. During the MENSTRUAL PHASE, the STRATUM FUNCTIONALIS of the endometrium is shed with a discharge of blood, tissue fluid, mucus, and epithelial cells.
2. OVARIAN FOLLICLES, called PRIMARY FOLLICLES, begin their development. At birth, each ovary contains about 200,000 such follicles, each consisting of a primary oocyte surrounded by a single flattened layer of epithelial cells.
3. During the early part of each menstrual phase, 20 to 25 primary follicles start to develop and produce low levels of estrogens. Toward the end of the menstrual phase (days 4 to 5), about 20, of the PRIMARY FOLLICLES develop into SECONDARY FOLLICLES, each of which consist of a secondary oocyte surrounded by several layers of epithelial cells.

K. FRC: PREOVULATORY PHASE

1. The PREOVULATORY PHASE is the time between menstruation and ovulation. This phase is more variable in duration than the other phases and lasts from 6 to 13 days in a 28-day cycle.
2. During this phase, ENDOMETRIAL TISSUE REPAIRS ITSELF and one of the secondary follicles develops into a mature, or GRAAFIAN FOLLICLE, ready for ovulation. This follicle produces a bulge on the surface of the ovary.
3. Functionally, ESTROGENS are the dominant ovarian hormones during this phase of the menstrual cycle.

L. FRC: OVULATION

1. OVULATION is the rupture of the GRAAFIAN FOLLICLE with release of the secondary oocyte into the pelvic cavity. This generally occurs on or about the 14th day of a 28-day cycle.
2. OVULATION is brought about by the INHIBITION OF FSH and a SURGE OF LH.
3. Signs of ovulation include INCREASED BASAL TEMPERATURE, CLEAR, STRETCHY CERVICAL MUCUS, CHANGES IN UTERINE CERVIX, and OVARIAN DISCOMFORT.
4. Following ovulation, the GRAAFIAN FOLLICLE collapses and the blood within it forms the CORPUS HEMORRHAGICUM.
5. This blood clot is eventually absorbed by the remaining follicular cells and enlarges to form the CORPUS LUTEUM. This occurs under the influence of LH.

M. FRC: POSTOVULATORY PHASE

1. The POSTOVULATORY PHASE is the most CONSTANT PHASE in the cycle and lasts from days 15 to 28 in a 28-day cycle. It represents the time between ovulation and the onset of the next menses. During this phase, the ENDOMETRIUM THICKENS in anticipation of implantation of a fertilized ovum. PROGESTERONE is the dominant hormone.
2. If fertilization and implantation do not occur, the CORPUS LUTEUM DEGENERATES and becomes the CORPUS ALBICANS. The decreased secretion of progesterone and estrogen by the degenerating corpus luteum then initiates another menstrual cycle.
3. IF FERTILIZED, the CORPUS LUTEUM is maintained by HUMAN CHORIONIC GONADOTROPIN (hCG) from the developing placenta. It also SECRETES ESTROGEN and PROGESTERONE to maintain pregnancy and to allow for lactation.
4. The onset of FEMALE PUBERTY is signaled by INCREASED LEVELS OF LH, FSH, and ESTROGEN. The FIRST MENSES, or MENARCHE, occurs at an average age of 12, preceded by initial development of the secondary sexual characteristics.

N. COMMON DISORDERS

1. SEXUALLY TRANSMITTED DISEASES (STDs) are spread by sexual contact and include gonorrhea, syphilis, genital herpes, chlamydia, trichomoniasis, and genital warts.
2. IMPOTENCE is the inability of the adult male to attain or hold an erection long enough to complete intercourse.
3. MALE INFERTILITY is the inability of the male's spermatozoa to fertilize the ovum.
4. MENSTRUAL DISORDERS include amenorrhea (absence of menstruation), dysmenorrhea (painful menstruation), abnormal bleeding, and premenstrual syndrome (PMS).
5. TOXIC SHOCK SYNDROME (TSS) includes widespread homeostatic imbalances and is a reaction to toxins produced by certain strains of Staplylococcus aureus. It is often associated with the use of tampons.
6. ENDOMETRIOSIS is a benign condition characterized by the growth of the uterine tissue outside of the uterus.
7. FEMALE INFERTILITY is the inability of the female to conceive.
8. PELVIC INFLAMMATORY DISEASE (PID) is a collective term used for any extensive bacterial infection of the pelvic organs, especially the uterus, Fallopian tubes, or ovaries.
9. Vulvovaginal candidiasis is a form of vaginitis caused by the yeast-like fungus Candida albicans.

IV. TEACHING TIPS AND SUGGESTIONS

A. Helpful Hints

1. It is helpful to identify some important medical procedures such as Pap smear, colposcopy, cone biopsy, and D&C.
2. Any information regarding breast cancer, breast palpation, and self-examination are very helpful, educational, and useful to all students.

B. Essays

1. Using a labeled diagram, explain the hormonal relationships that exist between the pituitary gland, ovarian cycle, and the menstrual cycle.
2. Trace a drop of sperm from its site of production to its release from the body. Include all major glands, ducts, tubes, and secretions.

V. AUDIOVISUAL MATERIALS

A. Overhead Transparencies

1. PAP Transparency Set (Trs. 28.1, 28.3a&c, 28.4, 28.6-28.14, 28.17-28.20 & 28.22-28.26).
2. Sex Education Set (16 Transparencies; 42 overlays; CARO).
3. Reproductive Systems- Unit II (11 Transparencies; RJB).

B. Videocassettes

1. Sharing the Future (26 min.; FHS).
2. Coming Together (26 min.; FHS).
3. The Biology of Human Sexuality (44 min.; CFH).
4. The Miracle of Life (60 min.; CARO).
5. Sexually Transmitted Diseases (28 min.; 1985; KSU).
6. Breast Self-Examination (15 min.; HRM).
7. Casual Encounters of the Infectious Kind (24 min.; 1974; EBEC/KSU).
8. Abortion: Public Issue or Private Matter? (25 min.; C; Sd; 1971; KSU).
9. About Contraception and Conception (11 min.; C; Sd; PE).
10. Babymakers (43 min.; C; Sd; 1979; KSU).
11. Birth Control: Five Effective Methods (10 min.; C; Sd; 1975; SEF).
12. Breast: Self-Examination (16 min.; C; Sd; ACS).
13. Ovulation (15 min.; C; Sd; UIFC).
14. V.D.: Very Communicable Diseases (19 min.; C; Sd; 1972; KSU).
15. V.D.: A Plague on Our House (35 min.; C; Sd; NBC).
16. When Life Begins (14 min.; C; Sd; KSU).
17. The Birth Control Movie (24 min.; C; Sd; PE).
18. Human Reproduction (21 min.; C; Sd; 1965; McG; KSU).
19. How Life Begins (46 min.; C; Sd; 1968; KSU).
20. Contraception (23 min.; C; Sd; 1973; KSU). ***Preview before showing**

C. Films: 16 mm

1. The Human Body: Reproductive System (16 min.; 1980; COR/KSU).
2. Meiosis (25 min.; UIFC).
3. Where Spermatozoa Are Formed (9 min.; UIFC).
4. Ovulation (15 min.; UIFC).
5. How Life Begins (46 min.; 1968; McG/KSU).
6. Human Reproduction (21 min.; 1965; McG/KSU).
7. Breast: Self-Examination (16 min.; ASC).
8. Your Pelvic and Breast Examination (12 min.; 1975; PE).
9. Vasectomy (17 min.; 1972; CHUR/KSU).
10. Birth Control: How? (32 min.; 1965; FI/KSU).

D. Transparencies: 35 mm (2x2)

1. PAP Slide Set (Slides 131-138).
2. AHA Slide Set.
3. Visual Approach to Histology: Male and Female Reproductive Systems (37 Slides; FAD).
4. Sex Is Not A Dirty Word (H&R).
5. Reproductive System and Its Function (EIL).
6. The Male and Female Reproductive Set (CARO).
7. Biology of Human Reproduction (40 Slides; CARO).

E. Computer Software

1. Animal Reproduction (Apple II; 39-8842; CARO).
2. Sexually Transmitted Diseases (Apple; C4154; EIL; SC-378011; PLP).
3. Contraception (Apple; C4329A; EIL; SC-378010; PLP).
4. Reproductive Systems (Apple; SC-182019; PLP).
5. Reproductive System Probe (Apple; SC-175046; PLP).
6. Reproductive Systems (Apple; SC-378013; PLP).
7. Sexually Transmitted Diseases (IBM; APPLE; MAC; PLP).
8. Graphic Human Anatomy and Physiology-Reproductive System (IBM; PLP).
9. Contraception (Apple; IBM; MAC; PLP).
10. Practicing Sexual Decision Making (Apple II; CARO).

CHAPTER 24

DEVELOPMENT AND INHERITANCE

CHAPTER AT A GLANCE

- SEXUAL INTERCOURSE
  - *Male Sexual Act*
    - *Erection*
    - *Lubrication*
    - *Orgasm*
  - *Female Sexual Act*
    - *Erection*
    - *Lubrication*
    - *Orgasm (Climax)*
- DEVELOPMENT DURING PREGNANCY
  - *Fertilization*
  - *Formation of the Morula*
  - *Development of the Blastocyst*
  - *Implantation*
  - *In Vitro Fertilization*
- EMBRYONIC DEVELOPMENT
  - *Beginnings of Organ Systems*
  - *Embryonic Membranes*
  - *Placenta and Umbilical Cord*
  - *Fetal Circulation*
- FETAL GROWTH
- HORMONES OF PREGNANCY
- GESTATION
- PRENATAL DIAGNOSTIC TECHNIQUES
  - *Amniocentesis*
  - *Chorionic Villus Sampling (CVS)*
- PARTURITION AND LABOR
- LACTATION
- BIRTH CONTROL
  - *Sterilization*
  - *Hormonal Methods*
  - *Intrauterine Devices (IUDs)*
  - *Barrier Methods*
  - *Chemical Methods*
  - *Physiologic (Natural) Methods*
  - *Coitus Interruptus (Withdrawal)*
  - *Induced Abortion*
- INHERITANCE
  - *Genotype and Phenotype*
  - *Genes and the Environment*

- *Inheritance of Sex*
- *Red-Green Color-blindness and Sex-linked Inheritance*
- MEDICAL TERMINOLOGY AND CONDITIONS
- WELLNESS FOCUS: EATING FOR TWO

I. CHAPTER SYNOPSIS

This chapter introduces a discussion of the male and female roles in sexual intercourse. This is followed by events of fertilization and the major developmental changes that occur up to and including implantation. Attention is given to embryonic development and fetal growth. Consideration is then given to the numerous hormones of pregnancy, the major changes during gestation, prenatal diagnostic techniques, and labor and parturition. Numerous methods of contraception are discussed. The chapter concludes with a general discussion of inheritance, medical terms and conditions.

II. LEARNING GOALS/STUDENT OBJECTIVES

1. Describe the roles of the male and female in sexual intercourse.
2. Explain how secondary oocyte is fertilized and implanted in the uterus.
3. Discuss alternative procedures to natural fertilization and implantation.
4. Describe the principal events associated with embryonic development.
5. Describe the principal events associated with fetal growth.
6. Explain the functions of the hormones secreted during pregnancy.
7. Describe the stages of labor.
8. Discuss how lactation occurs and how it is controlled.
9. Name and compare the effectiveness of birth control (BC) methods.
10. Describe the basic principles of inheritance.

III. SAMPLE LECTURE OUTLINE

A. SEXUAL INTERCOURSE

1. SEXUAL INTERCOURSE (copulation or coitus) is the process by which spermatozoa are deposited in the vagina.
2. The role of the male involves penile erection, minimal lubrication, and ejaculation, a component of orgasm.
3. The female role involves clitoral erection, significant lubrication, and orgasm.

B. DEVELOPMENT DURING PREGNANCY

1. PREGNANCY is a sequence of events that normally includes FERTILIZATION, IMPLANTATION, EMBRYO GROWTH, and FETAL GROWTH that terminates in birth. All events are hormonally controlled.
2. FERTILIZATION refers to the penetration of the secondary oocyte by a sperm cell and the subsequent union of the sperm and oocyte nuclei to form a zygote.

3. Fertilization normally occurs in the FALLOPIAN TUBES when the oocyte is approximately one-third of the way down the tube, usually within 24 hours of ovulation.
4. Although sperm undergo maturation in the epididymis, they are not able to fertilize until they have remained within the female reproductive tract for about 10 hours. This process is called CAPACITATION.
5. The PENETRATION of the oocyte by the sperm is facilitated by enzymes produced by the ACROSOME of the sperm cell, which penetrate the CORONA RADIATA and ZONA PELLUCIDA around the oocyte.
6. Normally one sperm fertilizes a secondary oocyte, producing a zygote after the secondary oocyte completes it equatorial division of meiosis II.
7. Immediately after fertilization, rapid cell division of the ZYGOTE occurs. This process is called CLEAVAGE. Successive cleavage produce a structure called the MORULA, a few days after fertilization.
8. As the number of cells in the morula increase, a ball of cells called the BLASTOCYST is produced. It is comprised of an outer layer, the TROPHOBLAST, surrounding a cavity, called the BLASTOCOEL, and an INNER CELL MASS, which will develop into the EMBRYO.
9. IMPLANTATION is the attachment of the BLASTOCYST to the endometrium. This is accomplished by enzymes secreted by the trophoblast that digest the uterine lining so that blastocyst can bury itself into the endometrium.
10. The PLACENTA will develop between the INNER CELL MASS and the ENDOMETRIAL WALL to provide nutrients for the growth of the embryo. At this time, the trophoblast begins to secrete HUMAN CHORIONIC GONADOTROPIN (hCG), a hormone that maintains the corpus luteum and the secretion of progesterone.

C. EMBRYONIC DEVELOPMENT

1. The first two months of development are considered the EMBRYONIC PERIOD, and during this time the developing human is called an EMBRYO.
2. The months of development after the second month are considered to be the FETAL PERIOD, and the developing human is called a FETUS.
3. By the end of the embryonic period, the rudiments of all principal adult organs are present, embryonic membranes are developed, and by the end of the third month, the placenta is formed.
4. The primary GERM LAYERS of ECTODERM, ENDODERM, and MESODERM form all tissues and organs of the developing organism.
5. The EMBRYONIC MEMBRANES, which lie outside of the embryo and protect and nourish the embryo, and later the fetus, include the AMNION, CHORION, and PLACENTA.
6. The AMNION is a thin, protective membrane that initially overlies the embryonic disc and is formed by the eighth day following fertilization. Through the course of development, it will surround the embryo, creating a cavity which becomes filled with amniotic fluid.
7. The AMNIOTIC FLUID is derived from a filtrate of maternal blood, and later the fetus will contribute to the amniotic fluid through urine excretion.
8. The amniotic fluid serves as a shock absorber for the fetus, assists in the regulation of fetal body temperature, and prevents adhesions between the skin of the fetus and the surrounding tissues.
9. The CHORION is derived from the ectoderm and mesoderm of the blastocyst. It surrounds the embryo and later, the fetus.

10. The CHORION is the principal embryonic part of the placenta, a structure through which materials are exchanged between mother and fetus.
11. The PLACENTA is developed by the third month of pregnancy and is formed by the chorion of the embryo and a portion of the endometrium of the mother.
12. The PLACENTA allows for the exchanges of oxygen, carbon dioxide, nutrients and metabolic wastes, and the storage of nutrients such as carbohydrates, proteins, calcium, and iron. These components are released into fetal circulation as needed.

D. FETAL CIRCULATION

1. FETAL CIRCULATION differs from adult circulation because the lungs, kidneys, and GI tract of the fetus are not functioning. The fetus derives oxygen and nutrients from maternal circulation via the placenta.
2. The exchange of all materials occurs through the PLACENTA, which is attached to the umbilicus of the fetus by an UMBILICAL CORD, and communicates with maternal circulation via numerous blood vessels which branch into the placenta.
3. Blood passes from the fetus to the placenta via two UMBILICAL ARTERIES in the umbilical cord. Exchanges occur and blood is returned from the placenta via a single UMBILICAL VEIN.
4. The UMBILICAL VEIN ascends to the LIVER and passes through a SHUNT called the DUCTUS VENOSUS, which by-passes the liver and empties into the inferior vena cava.
5. Blood entering the fetal right atrium bypasses the right ventricle via a shunt through the interatrial septum called the FORAMEN OVALE.
6. Blood reaching the right ventricle and pumped out of the pulmonary artery is shunted from the lungs through a connection between the pulmonary artery and aorta called the DUCTUS ARTERIOSUS. This will transport blood directly into the aorta to be sent to all parts of the body.

E. FETAL GROWTH

1. During the fetal period, organs established by the primary germ layers grow rapidly and the organism takes on a human appearance.
2. A complete summary of changes associated with the embryonic and fetal period is presented in Exhibit 24-2.

F. HORMONES OF PREGNANCY

1. The CORPUS LUTEUM is maintained for at least the first three to four months of pregnancy, during which time it continues to secrete estrogens and progesterone.
2. The CHORION of the placenta secretes human chorionic gonadotropin (hCG), which stimulates continued production of progesterone by the corpus luteum.
3. After the third or fourth month of pregnancy, the PLACENTA will secrete estrogens, and by the sixth week progesterone. The levels of hCG will then be greatly reduced because the secretions of the corpus luteum are no longer needed. These placental hormones will increase until the time of delivery.
4. HUMAN CHORIONIC SOMATOMAMMOTROPIN (hCS) is also secreted by the chorion and functions to stimulate development of the breast tissue for lactation.
5. RELAXIN is produced by the placenta and ovaries. It relaxes the symphysis pubis and helps to dilate the cervix.

6. INHIBIN is also produced by the ovaries and has been found in the human placenta. It functions to inhibit the secretion of FSH and may regulate the secretion of hCG.

G. GESTATION

1. The period of time from conception to birth is called GESTATION.
2. The human gestation period is approximately 280 days from the first day of the last menstrual period.
3. By the end of the third month of gestation, the uterus occupies most of the pelvic cavity. As the fetus continues to grow, the uterus extends higher into the abdominal cavity.
4. At TERM, the uterus practically fills the abdominopelvic cavity, causing displacement of the maternal intestines, liver, and stomach upward, elevation of the diaphragm, and a widening of the thoracic cavity. In the pelvic cavity, there is a compression of the ureters and urinary bladder.
5. By the 27th week of gestation, maternal cardiac output increases 20-30%, tidal volume increases, expiratory reserve volume decreases, and shortness of breath will continue for the last few months.

H. PRENATAL DIAGNOSTIC TECHNIQUES

1. AMNIOCENTESIS involves extracting and testing a sample of amniotic fluid and is used to diagnose genetic disorders and to determine fetal maturity and well-being. It can be performed no earlier than 14 to 16 weeks after conception.
2. CHORIONIC VILLUS SAMPLING (CVS) is used to detect prenatal genetic defects. It picks up the same defects as amniocentesis but can be performed as early as 8 to 10 weeks after conception. The procedure does not involve the penetration of the maternal abdominal wall, uterine wall, or amniotic cavity.

I. PARTURITION AND LABOR

1. The term PARTURITION refers to birth and is accompanied by a sequence of events called LABOR. The onset of labor is hormonally directed through the levels of estrogen, progesterone, and more importantly, oxytocin, which stimulates uterine contractions.
2. Labor can be divided into three phases: stage of DILATION, stage of EXPULSION, and PLACENTA STAGE.
3. The STAGE of DILATION is the time from the onset of labor to the complete dilation of the uterine cervix to a diameter of approximately 10 cm. The rupture of the amniotic sac usually takes place in this stage.
4. The STAGE of EXPULSION is the time from complete cervical dilation to delivery.
5. The PLACENTA STAGE is the time between the birth of the baby and the expulsion of the placenta by uterine contractions.

J. LACTATION

1. LACTATION is the secretion and ejection of milk by the mammary glands. The major hormone involved is prolactin (PRL).
2. PRL is released by the anterior pituitary in response to prolactin releasing factor (PRF).
3. The principal stimulus for lactation is the sucking action of the infant.

4. Sucking sends impulses from the nipples to the hypothalamus and PRF is released. The anterior pituitary is then stimulated to release PRL.
5. The sucking action also initiates impulses to the HYPOTHALAMUS for the release of OXYTOCIN by the posterior pituitary.
6. OXYTOCIN promotes milk letdown, the process by which milk is moved from the alveoli of the mammary glands into the ducts so it can be ejected.

K. BIRTH CONTROL

1. Several types of contraceptive methods are available. Each has its own advantages and disadvantages. The methods discussed are sterilization, hormones, intrauterine devices, physical barriers, chemical spermicides, physiological (natural) means, coitus interruptus (withdrawal), and induced abortion.
2. STERILIZATION involves vasectomy in males, and tubal ligation in females. The result is that neither sperm nor egg will be able to reach their normal destinations.
3. The hormonal method, THE PILL, prevents ovulation, and consequently fertilization. The most common birth control pills contain high concentration of progesterone and low concentrations of estrogen. These two hormones inhibit the release of FSH and LH by inhibiting GnRH by the hypothalamus.
4. An INTRAUTERINE DEVICE (IUD) is a small plastic, copper, or stainless steel object inserted into the cavity of the uterus. It is unclear how the IUD works, but it is believed that it causes changes in the uterine lining that produce a substance that either destroys sperm or the ova.
5. BARRIER METHODS are designed to prevent spermatozoa from gaining access to the uterine cavity and Fallopian tubes. They include condoms, cervical diaphragms, and cervical caps.
6. CHEMICAL METHODS of contraception refer to various foams, creams, jellies, suppositories, and douches that contain spermicidal agents, thus making the vagina and cervix unfavorable for sperm survival.
7. NATURAL OR PHYSIOLOGIC methods rely on the natural menstrual cycle of the female. This is generally called the rhythm method.
8. COITUS INTERRUPTUS refers to the withdrawal of the penis from the vagina prior to ejaculation.
9. ABORTION refers to the premature expulsion from the uterus of the products of conception.
10. A detailed summary of contraceptive methods, their advantages and disadvantages, and success rates are given in Exhibit 24-3.

L. INHERITANCE

1. INHERITANCE is the passage of hereditary traits from one generation to another.
2. The nuclei of all human cells except the gametes contain 23 pairs of chromosomes. One chromosome from each pair comes from the mother, the other from the father.
3. HOMOLOGOUS CHROMOSOMES contain genes that control the same traits.
4. The actual genetic makeup of an individual is referred to as the GENOTYPE; the physical or outward expression of a trait is called the PHENOTYPE.
5. Genes that dominate the phenotypic expression are called DOMINANT GENES. The gene that is not expressed is called a RECESSIVE GENE..
6. A comparison of dominant and recessive human traits is given in Exhibit 24-4.
7. The genotype of an individual is not simply a dominant-recessive interaction. Other factors in the environment can also be influencing factors.

8. A TERATOGEN is any agent or influence that causes physical defects in the developing embryo by changing the genotypes. Examples are defoliants, some industrial chemicals, thalidomide, diethylstilbestrol (DES), LSD, and ionizing radiation.

M. INHERITANCE OF SEX

1. One pair of chromosomes in cells differs in males and females. In females, there is a pair of SEX CHROMOSOMES, represented by XX. In the male, the sex chromosomes are represented by XY. The remaining 22 pairs are referred to as the autosomes.
2. The sex chromosomes, particularly the Y-chromosomes, are responsible for the sex of the individual.
3. A gene called the TESTIS-DETERMINING FACTOR (TDF) is contained in a fragment of the Y chromosome. The presence of TDF in a fertilized ovum directs the fetus to produce testes and differentiate into a male.
4. The sex chromosomes are responsible for the transmission of some non-sexual traits, and are referred to as sex-linked traits. Among these are COLOR-BLINDNESS, BALDNESS, and X-LINKED HEMOPHILIA.

IV. TEACHING TIPS AND SUGGESTIONS

A. HELPFUL HINTS

1. It is informative to discuss the basic process of in vitro fertilization (IVF) and gamete intrafallopian transfer (GIFT), as alternative methods for achieving pregnancy.
2. A discussion of prenatal surgery, such as diaphragmatic hernia repair, will aid in understanding the benefit of these advances in technology.
3. Recent evidence has shown that body fat has a regulatory role in reproduction. The entire reproductive process can be offset with too much or too little body fat.
4. A film on childbirth will provide the students with some knowledge of the events that occur.
5. Detail the differences between vaginal deliveries and cesarean delivery.
6. A discussion of several common genetic disorders such as Down's syndrome, Klinefelter's syndrome, and Turner's syndrome will illustrate how chromosomal dysfunctions occur.

B. ESSAYS

1. An X-linked trait is one that appears more frequently in one sex than another. Hemophilia is such as example and occurs more frequently in males. Explain how hemophilia can be inherited. Assume that a father is normal and that the mother is also normal, but carries the recessive gene for hemophilia.
2. Describe the hormones and the interaction needed for the maintenance of the endometrium and for lactation.
3. Outline the major developmental changes in fetal development and compare these with those that occur in embryonic development.

C. Topics for Discussion

1. Discuss the implications of IVF and GIFT on the future of the human society.
2. Discuss why certain genetic disorders, such as Down's syndrome, are not continually inherited in a family.

V. AUDIOVISUAL MATERIALS

A. Overhead Transparencies

1. PAP Transparency Set (Trs. 28.1, 28.3a&c, 28.4, 28.6, 28.7-28.14, 28.17-28.20, 28.22-28.26).
2. Genetics: Determination of Sex (GAF).
3. Sex Education Set (16 Transparencies; CARO).
4. Reproductive System (RJB).

B. Videocassettes

1. Sharing the Future (26 min.; FHS).
2. Coming Together (26 min.; FHS).
3. The Biology of Human Sexuality (44 min.; CFH).
4. The New Life (26 min.; FHS).
5. Into the World (26 min.; FHS).
6. Life in the Womb (40 min.; HRM).
7. The Gift of Love (30 min.; NDSS).
8. The Miracle of Life (60 min.; CARO).
9. Sexually Transmitted Diseases (28 min.; 1985; KSU).
10. Breast: Self-Examination (15 min.; HRM).
11. Casual Encounters of the Infectious Kind (24 min.; 1974; EBEC/KSU).
12. Abortion: Public Issue or Private Matter? (25 min.; C; Sd; 1971; KSU).
13. About Contraception and Conception (11 min.; C; Sd; PE).
14. Babymakers (43 min.; C; Sd; 1979; KSU).
15. Birth Control: Five Effective Methods (10 min.; C; Sd; 1975; SEF).
16. Breast: Self-Examination (16 min.; C; Sd; ACS).
17. Ovulation (15 min.; C; Sd; UIFC).
18. V.D.: Very Communicable Diseases (19 min.; C; Sd; 1972; KSU).
19. V.D.: A Plague on Our House (35 min.; C; Sd; NBC).
20. When Life Begins (14 min.; C; Sd; KSU).
21. The Birth Control Movie (24 min.; C; Sd; PE).
22. Human Reproduction (21 min.; C; Sd; 1965; McG; KSU).
23. How Life Begins (46 min.; C; Sd; 1968; KSU).
24. Contraception (23 min.; C; Sd; 1973; KSU). ***Preview before showing**

C. Films: 16 mm

1. The Human Body: Reproductive System (16 min.; 1980; COR/KSU).
2. How Life Begins (46 min.; 1968; McG/KSU).

3. Human Reproduction (21 min.; 1965; McG/KSU).
4. Breast: Self-Examination (16 min.; ASC).
5. Your Pelvic and Breast Examination (12 min.; 1975; PE).
6. Vasectomy (17 min.; 1972; CHUR/KSU).
7. Birth Control: How? (32 min.; 1965; FI/KSU).
8. Normal Birth (10 min.; KSU).
9. Gentle Birth: Leboyer (15 min.; 1976; NYF).
10. The Beginning of Life (30 min.; 1968; KSU).
11. Have a Healthy Baby (22 min.; 1978; CHUR/KSU).
12. Laws of Heredity (15 min.; EBF).
13. A Child Is A Child (7 min.; KSU).

D. TRANSPARENCIES: 35 MM (2X2)

1. PAP Slide Set (Slides 139-140).
2. AHA Slide Set.
3. Life in the Womb, Parts I-II (160 Slides; IBIS).
4. Genetics, I-II (219 Slides; BM).
5. The Genetic Material (59 Slides; BM).
6. Heredity, Health, and Genetic Disorders (1670 Slides; IBIS).
7. Neonatology Slide Series (218 Slides; CARO).

E. COMPUTER SOFTWARE

1. Pregnancy and Prenatal Baby Care (Apple; C4322A; EIL).
2. Birth Defects (Apple; C4323A; EIL).
3. Prenatal Development and Childbirth (Apple; SC-378012; PLP).
4. Pregnancy and Health (Apple; SC-390323; PLP).
5. The PAP Test (IBM; PLP).
6. The Reproductive System (IBM; PLP).
7. Menstruation (Apple; PLP).
8. Venereal Disease (Apple; IBM; MAC; PLP).
9. Contraception (Apple; IBM; MAC; EI).
10. Contraception (Apple; PLP).
11. Sexually Transmitted Diseases (IBM: Apple; MAC; PLP).